P9-BZV-551

THE CONFIDENCE CODE

ALSO BY KATTY KAY AND CLAIRE SHIPMAN

Womenomics

THE CONFIDENCE CODE

The Science and Art of Self-Assurance—What Women Should Know

Katty Kay

and Claire Shipman

An Imprint of HarperCollins*Publishers*
www.harpercollins.com

 For information, address HarperCollins Publishers, 10 East 53rd Street, New York, NY 10022.

HarperCollins books may be purchased for educational, business, or sales promotional use. For information, please e-mail the Special Markets Department at SPsales@harpercollins.com.

FIRST EDITION

Designed by Ruth Lee Mui

Library of Congress Cataloging-in-Publication Data has been applied for.

ISBN: 978-0-06-223062-1

14 15 16 17 18 DIX/RRD 10 9 8 7 6 5 4 3 2 1

For our daughters,

Maya, Poppy, and Della,

and our sons,

Felix, Jude, and Hugo

Contents

Introduction

There is a quality that sets some people apart. It is hard to define but easy to recognize. With it, you can take on the world; without it, you live stuck at the starting block of your potential.

There's no question that twenty-eight-year-old Susan had plenty of it. Like many of us, though, she was terrified of public speaking. Susan had a lot to say—she just didn't like the spotlight. She confessed to friends that she spent many sleepless nights worrying about upcoming performances, fearful of being ridiculed. Her early speaking efforts weren't great. But she kept at it. Armed with a sheaf of notes and protected by her sensible dresses, she fought her nerves and delivered her controversial message over and over, often to extremely skeptical male audiences. She knew she had to conquer her fear to do her job well. And she did, becoming a very persuasive public speaker indeed.

Susan B. Anthony, the voice of women's suffrage for the United States, worked for fifty years to win women the right to vote. She died in 1906, fourteen years too soon to see what she'd accomplished, but she was never deterred—either by her vulnerabilities, or by the fact that victory was always just out of reach.

Just making the trip to school every day, as a girl in modern-day Pakistan, requires that same quality. And then to imagine, as a twelve-year-old, that you can challenge the Taliban by calling for education reform, blogging to the world as schools are blown up around you, absolutely demands it. And it calls for a huge dose of something remarkable to keep going, to keep fighting for a cause, after being pulled off a bus, shot in the head by extremists, and left for dead at fourteen. Malala Yousafzai has courage, to be sure. When the Taliban announced they intended to kill her she barely seemed to blink, saying: "I think of it often and imagine the scene clearly. Even if they come to kill me, I will tell them what they are trying to do is wrong, that education is our basic right."

But she's harnessing something else, too, something that fuels her defiance and charts her steady movement forward. Malala nurtures an extraordinary, almost unimaginable belief that she can succeed, even when the odds are stacked, boulderlike, against her.

A century apart, these two women are united by a common faith—a sense that they can achieve what they set out to do. What they share is confidence. It's potent, essential even, and for women, it's in alarmingly short supply.

The elusive nature of confidence has intrigued us for years, ever since we started writing *Womenomics* in 2008. We were busy detailing the positive changes unfolding for women: remarkable data about our value to the bottom line of companies, and the power that gives us to balance our lives and still succeed. But as we talked to women, dozens of them, all accomplished and credentialed, we kept bumping up against a dark spot that we couldn't quite identify, a force clearly holding us back. Why did the successful investment banker mention to us that she didn't really deserve the big promotion she'd just got? What did it mean when the fast-rising engineer, who'd been a pioneering woman in her industry for decades, told us offhandedly that she wasn't sure she was really the best choice to run her firm's new big project?

In two decades of covering American politics, we have interviewed some of the most influential women in the nation. In our jobs and our lives, we walk among people who you'd assume would brim with confidence. On closer inspection, however, with our new focus, we were surprised to realize the full extent to which the power centers of this nation are zones of female self-doubt. Woman after woman, from lawmakers to CEOs, expressed to us some version of the same inexplicable feeling that they don't fully own their right to rule the top. Too many of the fantastically capable women we met and spoke with seemed to lack a certain boldness, a firm faith in their abilities. And for some powerful women, we discovered, the very subject is uncomfortable, because it might reveal what they believe to be an embarrassing weakness. If *they* are feeling all that, only imagine what it is like for the rest of us.

You know those uneasy sensations: the fear that if you speak out you will sound either stupid or self-aggrandizing; the sense that your success is unexpected and undeserved; the anxiety you have about leaving your comfort zone to try something exciting and hard and possibly risky.

We have often felt the same kind of hesitation ourselves. Comparing notes about confidence levels at the end of a dinner a few years ago, as well as we knew each other, was a revelation. Katty went to a top university, got a good degree, speaks several languages, and yet she has spent her life convinced she just isn't intelligent enough to compete for the most prestigious jobs in journalism. Claire found *that* implausible, laughable really, and yet realized that she too, for years, routinely deferred to the alpha-male journalists around her, assuming that because they were so much louder, so much more certain, that they just knew more. She almost unconsciously believed that they had a right to talk more on television. Were they really just more self-assured?

The questions kept coming. Had we merely stumbled across a few anecdotes here and there, or are women really less confident than men? And what is confidence, anyway? What does it let us do? How critical is it to our well-being? To success? Are we born with it? Can we get more of it? Are we creating it or thwarting it in our kids? Finding answers to these questions was clearly our next project.

We covered more territory than we initially envisioned, because each interview and each answer convinced us that confidence is not only an essential life ingredient, but also unexpectedly complex. We met with scientists who study the way confidence manifests itself in lab rats and monkeys. We talked to neurologists who suggested that it is rooted in our DNA, and psychologists who told us it is the product of the choices we make. We talked to coaches, of performance and sports, who told us it comes from hard work and practice. We tracked down women who clearly have it, and women who don't so much, to get their take. And we talked to men—bosses, friends, and spouses. Much of what we found is relevant for both sexes; our genetic blueprints aren't wildly different when it comes to confidence. But there is a particular crisis for women.

For years, we women have kept our heads down and played by the rules. We have made undeniable progress. Yet we still haven't reached the heights we know we are capable of scaling. Some misguided bigots suggest women aren't competent. (Personally, we haven't found many incompetent women.) Others say children change our priorities, and, yes, there is some truth in this claim. Our maternal instincts do indeed create a complicated emotional tug between our home and work lives that, at least for now, just doesn't exist as fiercely for most men. Many commentators point to cultural and institutional barriers aligned against us. There's truth in that, too, but all of these reasons are missing something more profound—our lack of self-belief.

We see it everywhere: Bright women with ideas to contribute who don't raise their hands in meetings. Passionate women who would make excellent leaders, but don't feel comfortable asking for votes or raising campaign money. Conscientious mothers who'd rather someone else become president of the PTA while they work behind the scenes. Why is it that women sound less sure of ourselves when we know we are right than men sound when they think they could be wrong?

Our complicated relationship with confidence is more pronounced in the workplace, in our public pursuits. But it can spill over to our home lives, undermining the very areas in which we have traditionally felt surer of ourselves. Think about it. You'd love to give a thoughtful toast at your best friend's birthday party, but even the prospect of speaking in front of thirty people makes you start to sweat—so you mutter a few words, keep it very short, and nurse a dissatisfied feeling that you haven't done her justice. You always wished you'd run for class president in college, but asking other people to vote for you, well, it just seemed so arrogant. Your brother-in-law is so annoying with his sexist views, but you're worried that if you stand up to him in front of everyone you'll come across as strident, and, anyway, he always seems so on top of his facts.

Imagine all the things over the years you wish you had said

or done or tried—but didn't because something held you back. Chances are, that something was a lack of confidence. Without it we are mired in unfulfilled desires, running excuses around in our heads, until we are paralyzed. It can be exhausting, frustrating, and depressing. Whether you work or you don't, whether you want the top job or the part-time job—wouldn't it just be great to slough off the anxiety and the fretting about all the things you'd love to try but don't trust yourself to do?

In the most basic terms, what we need to do is start acting and risking and failing, and stop mumbling and apologizing and prevaricating. It isn't that women don't have the ability to succeed; it's that we don't seem to believe we *can* succeed, and that stops us from even trying. Women are so keen to get everything just right that we are terrified of getting something wrong. But, if we don't take risks, we'll never reach the next level.

The thoroughly accomplished twenty-first-century woman should spend less time worrying about whether she's competent enough and more time focused on self-belief and action. Competence she has plenty of.

The *Economist* magazine recently called female economic empowerment the most profound social change of our times. Women in the United States now get more college and graduate degrees than men do. We run some of the greatest companies. There are seventeen female heads of state around the world. We control more than 80 percent of U.S. consumer spending and, by 2018, wives will outearn husbands in the United States. Now comprising half of the workforce, women are closing the gap in middle management. Our competence and ability to excel have never been more obvious. Those who follow society's shifting values with a precision lens see a world moving in a female direction.

And yet.

At the top, our numbers are still small and barely increasing. On all levels, our talents are not being fully realized. We believe we're stalling because, all too often, women don't see, can't even envision, what's possible.

"When a man, imagining his future career, looks in the mirror, he sees a senator staring back. A woman would never be so presumptuous." That disarmingly simple observation from Marie Wilson, a veteran of women's political movements, was in many ways the launchpad for this exploration. It rang so true to us because it perfectly encapsulates both our reticence and our insecurity. And we'd add to it. Even when we *are* senators or CEOs or top performers of some sort, we don't recognize ourselves and our triumphs in the mirror. Women who have reached admirable heights have not always erased the nagging feeling that they might be unmasked as incompetent pretenders. And rather than diminishing with success, that feeling often grows the higher we climb.

A year before her book *Lean In* was published, Facebook COO Sheryl Sandberg told us "there are still days I wake up feeling like a fraud, not sure I should be where I am." Likewise the two of us spent years attributing our own success to luck, or, like Blanche DuBois, to the kindness of strangers. And we weren't being deliberately self-deprecating—we actually *believed* it. After all, how could we possibly have deserved to get to where we'd gotten?

Often women's self-assurance dwindles in more prosaic patterns. Peggy McIntosh, a sociologist at Wellesley College who has written extensively on what is aptly called the fraud syndrome, vividly remembers a conference she attended: "Seventeen women in a row spoke during the plenary session, and all seventeen started their remarks with some sort of apology or disclaimer. 'I just have one point to make,' or 'I've never thought about this very much' or 'I really don't know whether this is accurate.' And it was a women's *leadership* conference!"

The data is pretty grim. Compared with men, we don't consider ourselves ready for promotions, we predict we'll do worse on tests, we flat out tell researchers in big numbers that we just don't feel confident at our jobs.

Part of the problem is we can't make sense of the rules. Women have long believed that if we just work harder and don't cause any bother, our natural talents will shine through and be rewarded. But then we have watched as the men around us get promoted over us and paid more, too. We know, deep down, they are no more capable than we are; indeed, often they are less so, but they project a level of comfort with themselves that gets them noticed and rewarded. That comfort, that self-assurance—it's confidence, or at least their version of it.

More often than not, the way confidence manifests itself in men is wholly unappealing and downright foreign to women. Most women aren't comfortable dominating conversations, throwing their weight around in a conference room, interrupting others, or touting their achievements. Some of us have tried these tactics over the years, only to find that it just isn't our style.

We should pause and say that we know when we talk about women en masse we are oversimplifying. Some women have already cracked this code and others, of course, won't always recognize themselves in these pages. We are far from monolithic as a gender. The subject is important enough to most women—women of every personality type, ethnic and religious background and income level, that we hope you will forgive our choice to occasionally generalize rather than constantly qualify. We're determined to cast deep and wide, because the subject merits it.

The stakes are too high to give up on finding confidence just because the prevailing masculine model might not fit, or the reality looks foreboding. There are too many opportunities we are missing out on. As we dissected academic papers and reviewed interview transcripts we decided that what we need is a blueprint for confidence, a confidence code, if you will, that will get women headed in the right direction.

Consider women like our friend Vanessa, who is a very successful fundraiser for a nonprofit organization. Recently, she was called in by the president of her organization for an annual review. She'd raised a lot of money for the group and assumed she was heading in for a serious pat on the back. Instead, he gave her a reality check. Yes, she'd done well bringing in funds, but if she ever wanted to be a senior leader in the organization, she needed to start making decisions. "It doesn't matter if they're right," he told her. "Your team just needs to know you can make a call and stick to it." Vanessa couldn't believe what she was hearing. *It doesn't matter if they're right?* That was simply anathema to her.

Yet Vanessa recognized the truth in what her boss had said. She was so focused on being perfect, and absolutely right, that she held back from making decisions, particularly decisions that needed to be made fast. Like so many women, Vanessa is a perfectionist, but that quest for perfection and those fourteen-hour workdays weren't what her group really needed. And moreover, her habits stopped her from taking the decisive action that was required.

If you ask scientists and academics, as we did, how optimism is defined, you'll get a fairly consistent answer. The same goes for happiness and many other basic psychological qualities; they've been dissected and examined so often and for so long that we now have a wealth of practical advice about cultivating these attributes in ourselves and in others. But the same doesn't hold, we discovered, for confidence. It is altogether a more enigmatic quality, and what we learned about it is not at all what we expected when we set out to discover its nature.

For one thing, there's a difference between bravado and confidence. We also came to see that confidence isn't all in your mind, and it isn't generated by exercises to boost self-esteem, either. Perhaps most striking of all, we found that success correlates more closely with confidence than it does with competence. Yes, there is evidence that confidence is *more important* than ability when it comes to getting ahead. This came as particularly unsettling news to us, having spent our own lives striving toward competence.

Another disturbing finding is that some of us are simply born with more confidence than others. It is, it turns out, partly genetic. We did our own genetic tests to see how we stack up. We'll share them with you later, but suffice it to say that we were surprised by the results. And we discovered that male and female brains do indeed work differently in ways that affect our self-assurance. Yes, that fact is controversial. Yes, it's also true.

Confidence is only part science, however. The other part is art. And how people live their lives ends up having a surprisingly big impact on their original confidence framework. The newest research shows that we can literally change our brains in ways that affect our thoughts and behavior at any age. And so, fortunately, a substantial part of the confidence code is what psychologists call *volitional*: our choice. With diligent effort, we can all choose to expand our confidence. But we will get there only if we stop trying to be perfect and start being prepared to fail.

What the scientists call *plasticity*, we call *hope*. If you work at it, you can indeed make your brain structure more confidence-prone. One thing we know about women is that we're never afraid of hard work.

As reporters, we've been lucky enough to explore the power corridors of the world looking for stories, and we've seen the possibilities that confidence gives a person. We notice how some people aim high, simply assuming they will succeed, while others spend the same time and energy thinking of dozens of reasons why they can't. As mothers, we've watched the impact confidence has on our children. We see the kids who are liberated to say and do and risk, learning as they go, and stockpiling lessons for the future. And we see the youngsters who hold themselves back, fearful of some unformed, undeserved consequence.

And as women, particularly thanks to this project, we have both *felt* the life-changing impact of confidence, in our professional and our personal lives. Indeed we've discovered that accomplishment is not its most meaningful measure. There's a singular sense of fulfillment you get from simply having it and putting it to good use. One scientist we interviewed described her occasional brushes with confidence in particularly resonant terms: "I feel a spectacular kind of lock-and-key relationship with the world," she told us. "I can achieve. And I'm connected." Life on confidence can be a remarkable thing.

1

IT'S NOT ENOUGH TO BE GOOD

Even before we found the door, we could hear, and feel, the pounding and thumping and barked direction echoing down the halls. We had come to the bowels of Washington, DC's massive sports complex, the Verizon Center, on a hunt for raw confidence. We wanted to see it in action, to watch it on the basketball court, where, we surmised, confidence must be impervious to the turbulence of ordinary life, untroubled by gender battles, and stripped down to its essence. We were looking for that "aha" moment, a depiction of confidence so clear and compelling that it would shake up our female psychological GPS and shout, "*This way. This is what your destination looks like. This is where you go.*"

It was the opening practice of the Washington Mystics' 2013 season, and the first thing we noticed, as we stepped inside the gritty, basement practice court, was the towering physique of the women. It wasn't simply that they soared, on average, to above six feet tall and possessed muscled arms we could only dream of. There was an air of command about them that came from having mastered one of the most aggressive and challenging professional sports that women can play.

Tracking down unadulturated confidence isn't easy. We'd seen plenty of stuff like it demonstrated in boardrooms, in political offices, and on factory floors. But often that confidence seemed fleeting, or warped by social dictates. Sometimes it just felt phony; a well-crafted performance hiding deep wells of self-doubt. We'd figured that sports would be somehow different. You can't fake confidence on the 94-by-50-foot polished floors of a professional basketball court. To win here you have to believe in yourself. No doubting, no debating, no dithering. As with all top athletic pursuits, excellence is precisely measured, chronicled, and judged. And, assuming the basic physical tools are there, the central ingredient for success in competitive sports is confidence. Legions of sports psychologists have testified to its elemental importance to the game. If it weren't so, and its deficit weren't a problem, there wouldn't be sports psychologists in the first place, right?

That's why we knew women's basketball would be a rich laboratory for us. Moreover, this particular petri dish is one of the few in which grown women can be observed working together mostly in isolation from men, which takes away a major confidence inhibitor.

There was plenty of action and intensity on the court that morning. The Mystics were trying to fight their way back from the worst two seasons in the WNBA's seventeen-year history. We were watching two players in particular. Monique Currie, or Mo, as her teammates call her, is a DC native—a prep school, and then Duke, basketball phenom. She's a star forward for the team, and the toughest player we saw. Her shoulders are strikingly broad, even for her six-foot frame, and they took on a determined curve as she attacked the basket again and again.

Crystal Langhorne, at six feet two inches, is a power forward. When she was in high school, her devout father had to be persuaded to let her play ball on Sundays. As a pro, she's gone from mediocre rookie to all-star player, with a lucrative Under Armour endorsement deal. A white headband held back her long dark hair as she glided toward the basket, shooting with Zen-like ease.

We'd only been there a few moments when an intense scrimmage started and there it was: a performance so fierce it fueled a dazzling blur of perfectly timed passing, artful fakes, and three-point shots. It was a startling show of agility and power.

Confidence is the purity of action produced by a mind free of doubt. That's how one of our experts defines it. And that's what we'd just seen on the court, we thought in triumph.

After practice, though, we found something else. When we sat down to talk with Monique and Crystal, our perfect snapshot blurred with a multitude of doubts and contradictions. Not even here, in the WNBA, had they quite broken the confidence barrier.

Without the court as a backdrop and out of their sleek athletic gear, Monique and Crystal looked somewhat less intimidating. Now they were just exceptionally tall, attractive young women, who, visibly drained, sank with relief into plush armchairs in the VIP room. Monique, who'd changed into a slim jean jacket and T-shirt, immediately became intense and engaged on the subject of confidence. We got the sense it came up a lot.

"Sometimes as players you can kind of struggle with your confidence," said Monique, "because things might not be going well, because you think you're not playing as well as you can. But to be playing at this level you have to believe in what you can do and you have to believe in your ability."

Crystal nodded, her face partially obscured by a Yankees cap. Then she joined in, noting that there's plenty that messes with female players' confidence that doesn't seem to affect the men. "Let's say I have a bad game," she suggested. "I'll think, 'Oh my gosh, we lost' and I'll feel like I really wanted to help the team win, and win for the fans. With guys, if they had a bad game, they're thinking, 'I had a bad game.' They shrug off the loss more quickly."

What was striking in talking to Crystal and Monique was that with every answer a comparison to the guys popped up, even before we asked about it. And the Mystics don't even compete directly with men. Indeed, the frustrations sounded so familiar that we could have been having this conversation with a group of women in our line of work. Why do men usually just assume they're so great? Why do mistakes and backhanded comments just seem to slide off them?

"On the court, it's kind of hard to say certain things or play rough," Crystal said, "because women get hurt feelings. Our assistant coach says guys just curse each other out and then forget about it."

"Not me," noted Monique, with a wry grin. "I'm a mean player."

"Mo is different—she is more like a male athlete," Crystal said, laughing in agreement. "You could say something to Mo and she would brush it off. She can yell. I've played with Mo for a while, so I know how she is."

Still, even Monique rolled her eyes when asked whether her wellspring of confidence is really as deep as that of the men. "For guys," said Monique, in the slightly mystified, irritated tone that we'd come to recognize, "I think they have maybe thirteen- or fifteen-player rosters, but all the way down to the last player on the bench, who doesn't get to play a single minute, I feel like his confidence and his ego is just as big as the player's who is the superstar of the team." She smiled, shook her head, and went on. "For women it's not like that. If you're not playing, or if you're not considered one of the better players on the team, I think it really messes with our confidence."

We wondered what the Mystics coach, whom we'd noticed during practice, thought about all of this. A half foot shorter and twice the age of most of the players, Mike Thibault, clad in a navy team polo, was one of the few men on the court. A legendary WNBA coach who brought years of victories to the rival Connecticut Sun, he'd just arrived in Washington with the mission of helping to turn the Mystics' fortunes around. He was in a unique position to talk about the subject of confidence in male and female athletes, having trained both. As an NBA scout, Thibault helped to recruit Michael Jordan. He became an assistant coach for the Los Angeles Lakers and has spent the last ten years coaching women. The propensity to dwell on failure and mistakes, and an inability to shut out the outside world are, in his mind, the biggest psychological impediments for his female players, and they directly affect performance and confidence on the court.

"There's probably a distinction between being tough on themselves and too judgmental," he said. "The best male players I've coached, whether it's Jordan or people like that, they are tough on themselves. They push themselves. But they also have an ability to get restarted more quickly. They don't let setbacks linger as long. And the women can."

"It's very hard for me because sometimes I kind of hold on to things longer than I should," agreed Mo. "I can get down on myself about missing a shot, even though I know I worked for it—it's still an adjustment to say, 'All right, just let the play go—move on to the next play.' Even at thirty, and after eight seasons at WNBA, that's something that I still have to work on."

"I feel like with women, you still want to please people," sighed Crystal. "I feel like that's what happened to me last year, in my playing. That's my problem, sometimes I just want to please people."

Mo shrugged. "If you have a male attitude and that type of swagger and confidence in yourself, you play better."

Honestly, none of this was what we had been expecting or hoping to hear. How . . . *messy*, that even in our perfectly imagined habitat of female basketball stars, the essence of confidence was still elusive—or at least still battered by the same familiar forces. Monique and Crystal had looked so . . . purely confident out there on the court. But after thirty minutes of talk we'd uncovered overthinking, people pleasing, and an inability to let go of defeats—three traits we had already realized belonged on a confidence blacklist.

If clean confidence couldn't be found in professional sports, where was it? We decided to explore a realm in which women are routinely pushed well beyond their comfort zones, in direct competition with men.

Officer Michaela Bilotta had just graduated with honors from the U.S. Naval Academy at Annapolis and was one of fourteen members of her class chosen to join the prestigious Explosive Ordnance Disposal (EOD) team. The EOD is responsible for dealing with and deactivating chemical, biological, and nuclear weapons in areas of conflict, and its members routinely deploy with Special Operations Forces. To be chosen, you have to be the best. When we congratulated Bilotta on her new post she immediately deflected our praise, calling it "part chance." We pointed out how she had just, unwittingly, refused to own her achievement. She offered a half smile.

"I think it definitely took me longer than it would have for some other people to admit that I was worthy of it," Bilotta confessed. "Even though from the outside, I can look in and think, you did all the work and you earned your spot." She paused. We were sitting with her in her parents' basement, which, we noticed, was overflowing with sports gear, trophies, and academic plaques—the souvenirs of raising five determined girls. No clues that would have suggested a childhood that didn't nurture self-belief. "I just doubted it," she said, shaking her head. "I wondered, 'how did this happen? I got so lucky.'"

Luck. What could be more divorced from luck than passing all of the clearly defined, objectively measured physical and mental and intellectual hurdles that the military neatly lays out for someone like Michaela Bilotta? How was it that she couldn't see that what she had accomplished wasn't just a fluke?

Of course, we know exactly how she feels. We too have been masters at attributing our successes to the vagaries of fate. Katty still entertains the notion that her public profile in America is thanks to her English accent, which must, she suspects, give her a few extra IQ points every time she opens her mouth. Claire spent years telling people she was "just lucky"—in the right place at the right time—when asked how she became a CNN correspondent covering the collapse of communism in Moscow when she was still in her twenties.

"For years I really believed that it *was* all luck. Even as I write this I have to fight that urge. What I've realized only recently is that by refusing to take credit for what I had achieved, I wasn't nurturing the confidence I needed for my *next* career steps," she admits. "I was literally quaking when it came time for me to go back to Washington and cover the White House. At the time, I thought to myself,

'I'll never learn to report on politics. I don't know anything about it.'" Preoccupied and insecure about whether she would measure up, she should have relied on what she had already achieved to give her a psychological boost.

The more we surveyed the landscape looking for pockets of flourishing confidence, the more we uncovered evidence of a shortage. The confidence gap is a chasm, stretching across professions, income levels, and generations, showing up in many guises, and in places where you least expect it.

At a conference we moderated at the State Department, former U.S. secretary of state Hillary Clinton spoke openly about the fear she felt when she decided to run for the Senate in 2000, after eight years as First Lady, decades as a political spouse, and a successful legal career. "It's hard to face public failure. I realized I was scared to lose," she told us. That caught us off guard. "I was finally pushed," she said, "by a high school women's basketball coach, who told me, 'Sure, you might lose. So what? Dare to compete, Mrs. Clinton. Dare to compete.'"

Elaine Chao dared to compete. She was the country's first Chinese-American cabinet secretary. For eight years, she served President George W. Bush as labor secretary, the only cabinet member to stay with the president through his entire administration. There wasn't much in her background to predict such lofty heights. Chao was born in Taiwan; she came to the United States on a freight ship at the age of eight, after her father finally managed to get together enough money to pay the fare. Her rise reads like a classic tale of hard work, risk, and ironclad confidence.

But when we asked Chao if she ever doubted her abilities during those years in office, she was wonderfully candid, and funny. "Constantly," she replied. "I'm Asian American, are you kidding?

My fear was that the newspapers would have blaring headlines like: 'Elaine Chao Failed, Disgraced Whole Family.'"

We'd figured, and hoped, that the younger generation might have markedly different tales to tell. But their stories are eerily similar. It is hard to imagine a more successful Gen-Y'er than Clara Shih, for example. The thirty-one-year-old tech entrepreneur founded the successful social media company Hearsay Social in 2010. She joined the board of Starbucks at the tender age of twenty-nine. She's one of the few female CEOs in the still techno-macho world of Silicon Valley. Although she hasn't let the confidence gap stop her from racking up a string of impressive achievements, even she admits that it has tripped her up. "At Stanford, I found the computer science major very difficult. I really had to work hard, especially in the upper-level courses," Shih told us. "Yet, somehow, I was convinced that others found it easy. At times I felt like an imposter." Shih even considered dropping out and switching to an easier major. On graduation day, she was astonished to learn she'd finished first in her class.

"I realized I had deserved to be there all along and that some of the geeky guys who talked a big game weren't necessarily smarter."

Tia Cudahy, a Washington, DC, lawyer who always appears calm, upbeat, and utterly in charge, told us that she'd recently formed a partnership with a colleague to do some outside consulting, something she'd long wanted to try. Lo and behold, they got a contract right away. "In my mind, though, I immediately jumped to what I couldn't do—what parts of the job I didn't feel qualified for," she told us. She almost turned down the contract but managed to battle her doubts.

We were talking to Tia over drinks, thankfully, because we could at least laugh—after we had sighed in recognition. It's all such a waste of time and energy, these bouts of self-doubt we all engage in. Why do we do it?

Confidence Served with Crème Brûlée

The six-foot-tall, silver-haired woman headed toward us in a Washington, DC, restaurant, wearing a refined dark tweed dress, radiated self-assurance with a distinct flair. As she walked through the chic restaurant, heads swiveled as diners recognized one of the most powerful women in the world. Christine Lagarde runs the International Monetary Fund, the 188-country organization whose mission is to stabilize the world's financial systems, loan money to selected countries, and force reform on those that need it. Suffice it to say, she'd been busy.

Since we conceived of this project, we imagined Lagarde as one of the best possible guides through this confidence thicket. She claims a powerful position in the almost all-male club of global finance titans and she uses her formidable profile to pressure companies and heads of state to get women to the top—not because it's politically correct but because, she believes, it is good for the health of the world economy. She makes the same arguments we did in *Womenomics*—diversity helps the bottom line.

Fittingly, perhaps, she got her current job because, as she was helping to stave off the global financial meltdown from her post as the French finance minister, her IMF predecessor Dominique Strauss-Kahn, who was also on the short list to run for president of France, was discovered to have been routinely cheating on his glamorous and successful wife. One might think such a résumé should make him even *more* qualified to be a French president. But the womanizing included allegations that he had sexually assaulted a hotel maid and a journalist. The case involving the maid was eventually dropped, but the scandal was a front-page story in the United States, and it eventually caught on across the Atlantic. It turns out that there is a line you can cross in France on the issue of sex.

Lagarde was deemed the best person to set things right. Her weapons—a rational, conciliatory style and straightforward smarts—helped to calm the testosterone-fueled international economic crisis and quell internal politics at the IMF.

When we met Lagarde in person, we were struck by her regal bearing and the chic, thick white hair framing her face in a feminine but not fussy way. (Her only flourish was a subtly patterned silk scarf, draped around her neck in an elegant style neither of us had ever seen, and certainly could never master. Infuriatingly French.) She introduced herself with a friendly, piercing look, and then smiled. She was charming and open as she told us about her two grown-up sons, her preference for biking rather than driving around Washington, and her long-distance French boyfriend.

Raised and educated in France, she spent a year after high school as an intern at the U.S. Capitol. After law school in Paris, she decided to return to the States after a boss at a French law firm told her that, as a woman, she would never make partner there. In fifteen years, she'd not only become a partner at Baker & McKenzie, a top Chicago-based international law firm, but also its first female chairman.

Over grilled trout and wilted spinach, she recalled plenty of self-doubt as she worked her way up the ladder. "I would often get nervous about presentations or speaking, and there were moments when I had to screw up my courage to raise my hand or make a point, rather than hanging back."

And what's more, this woman who sits down in meetings next

to some of the most powerful men in the world and proceeds to tell them they need to change their ways and run their economies differently *still* worries about being caught off guard. "There are moments where I have to sort of go deep inside myself and pull my strength, confidence, background, history, experience and all the rest of it, to assert a particular point."

To compensate, we learned over the course of the meal, Lagarde zealously overprepares for everything. And with whom does she commiserate about how difficult that is? One of the few women at her altitude—the German chancellor.

"Angela Merkel and I have talked about it," she confided. "We have discovered that we both have the same habit. When we work on a particular matter, we will work the file inside, outside, sideways, backwards, historically, genetically and geographically. We want to be completely on top of everything and we want to understand it all and we don't want to be fooled by somebody else."

We pushed aside the crème brûlée we'd been enjoying and paused, taking in the image of *two* of the most powerful women in the world huddling someplace to compare notes about their mutual need to overprepare. Lagarde then volunteered something most men never would: "We assume, somehow, that we don't have the level of expertise to be able to grasp the whole thing."

"Of course it is part of the confidence issue," she concluded, shrugging, "to be overly prepared and to be rehearsed, and to make sure that you are going to get it all and not make a mistake." Is it a problem? we asked. "Well," she joked, "it's very time consuming!"

Perfectionism was very much on our growing list of confidence killers and so our role model had impressed us with her self-assurance, yet also managed to make a compelling case for the depth of the problem. (Like misery, we found ourselves perversely comforted to have more company. If Amazonian athletes, hard-charging military graduates, and global financiers are susceptible to self-doubt, no wonder we mere mortals have issues.) In all though, despite her vulnerabilities, Lagarde came across as confident, and confident in a way we'd like to be, and we would think about that contradiction for months.

The evening we had dinner with her, she was just back from the World Economic Forum's annual meeting in Davos, Switzerland, and laughingly recalled a panel about women's economic progress on which she had participated with other female luminaries, including her friend Sheryl Sandberg.

"So there we were, with just one token man right in the middle, poor guy. He struggled quite a bit, trying to be pushy. We were trying to listen to the moderator, or signal to her when we wanted to join the debate. He couldn't care less. He ignored her and talked when he liked. And in that setting, he came across as quite rude."

The incident had reminded her that women don't necessarily have to compete according to the guidelines of a male playbook.

"To the extent that it is more interesting to be female than male, why would we have to repress that rather than be ourselves with strength and worthiness? I've always said that we should not try to imitate the boys in everything they do."

It was an interesting thought, but one we wouldn't fully appreciate until later.

20 Percent Less Valuable

The shortage of female confidence is more than just a collection of high-octane anecdotes or wrenchingly familiar scenarios. It is increasingly well quantified and documented. The Institute of

Leadership and Management, in the United Kingdom, conducted a study in 2011, simply asking British women, in a series of questions, how confident they feel in their professions. Not very, as it turns out. Half of the women reported feelings of self-doubt about their performance and careers, while less than a third of male respondents reported self-doubt.

Linda Babcock, a professor of economics at Carnegie Mellon University and the author of *Women Don't Ask*, has uncovered a similar lack of confidence among American women, with concrete consequences. She found, in studies with business school students, that men initiate salary negotiations four times as often as women, and that when women do negotiate, they ask for thirty percent less than men do.

At Manchester Business School in England, Professor Marilyn Davidson says the problem stems from a lack of confidence and expectation. Each year she asks her students what they expect to earn, and what they deserve to earn, five years after graduation. "I've been doing this for about seven years," she said, "and every year there are massive differences between the male and female responses. The male students *expect* to earn significantly more than the women, and when you look at what the students think they *deserve* to earn, again the differences are huge." On average, she says, the men think they deserve $80,000 a year and the women $64,000—a $16,000 difference.

As reporters we are always thrilled to find precise measures, but still, the number is dismaying. Think about that for a minute. What Davidson's findings really mean is that women effectively believe they are 20 percent less valuable than men believe they are.

A more meticulous study by Cornell psychologist David Dunning and University of Washington psychologist Joyce Ehrlinger homed in on the perplexing issue of female confidence versus competence. At the time, Dunning and a Cornell colleague, Justin Kruger, were just finishing their seminal work on something called the Dunning-Kruger effect—the tendency for some people to substantially overestimate their abilities. (The less competent they are, the more they overestimate their abilities. Think about it for a minute. It makes strange sense.)

Dunning and Ehrlinger wanted to focus specifically on women, and the impact of their preconceived notions about their ability on their confidence. They gave male and female college students a pop quiz on scientific reasoning. Before the quiz, the students rated themselves on their scientific skills. "We wanted to see whether your general perception of 'Am I good in science?' shapes your impression of something that should be separate: 'Did I get this question right?'" Ehrlinger said. The women rated themselves more negatively than the men did in scientific ability. On a scale of 1 to 10, women gave themselves a 6.5 on average, and men gave themselves a 7.6. When it came to assessing how well they answered the questions, women thought they got 5.8 out of 10 right, men 7.1. And how did they actually perform? Their average was almost the same—women got 7.5 out of 10 and men 7.9.

In a final layer, to show the real impact of self-perception, the students were then asked, having no knowledge about how they'd performed, to participate in a science competition for prizes. The women were much more likely to turn down that opportunity—only 49 percent signed up for the competition, compared with 71 percent of the men.

"That was a proxy for whether women might seek out certain opportunities," said Ehrlinger. "Because they are less confident in general in their abilities, that led them to be less confident when they are actually performing in an achievement-related task. This then led them not to want to pursue future opportunities." It was a concrete example, in other words, of the real-world results of a lack of confidence.

The data confirms what we instinctively already know. Another example: We know that most women tend to talk less when we're outnumbered. We go into a meeting, study the layout, and choose to sit at the back of the room. We often keep our thoughts, which we decide can't be all that impressive, to ourselves. Then we get cross with ourselves when the male colleague next to us sounds smart saying the same thing that we would have said.

One Princeton research team set out to measure how much less women talk. Male and female volunteers were put to work solving a budget challenge. The study found that in some cases women, when in the minority, spoke 75 percent less than men did. Do we believe our words are that much less valuable? Do we think they are just as valuable, but we don't have the nerve to spit them out? Either way, we're underselling ourselves. The kicker is that a man in a room with mostly women talks just as much as he always does.

"It's so frustrating that we are typically so silent," says Virginia Shore, the chief curator of the State Department's Office of Art in Embassies and a leading expert on international contemporary art. "I certainly think of myself as confident. In my office, I'm a warrior, and I feel extremely comfortable in the world of art. But, when I step out of my office to go to weekly conferences at the State Department, it changes dramatically. It's all men around the table. Usually thirty men, and maybe a few women." She seemed comforted to hear that the research shows meetings just like hers happen everywhere.

Brenda Major, a social psychologist at the University of California in Santa Barbara, started studying the problem of self-perception decades ago. "In my earliest days as a young professor, I was doing a lot of work on gender, and I would set up a test where I'd ask men and women how they thought they were going to do on a variety of tasks or tests." She found that the men consistently overestimated their abilities and subsequent performance, and that the women routinely underestimated both. The actual performances did not differ in quality.

"It is one of the most consistent findings you can have," Major says of that test. And still, today, when she wants to suggest a study to her students where the results are utterly predictable, she points to this one.

On the other side of the country, the same thing plays out every day in Victoria Brescoll's lecture hall at Yale's School of Management. MBA students are nurtured specifically to project confidence in the fashion demanded by today's business world. While she sees from their performance that all of her students are top-of-the-chart smart, she's been startled to uncover her female students' lack of belief in themselves.

"There's just a natural sort of feeling among the women that they will not get a prestigious job, so why bother trying," she explained. "Or they think that they are not totally competent in the area, so they're not going to go for it."

What often happens to the female students is that they opt out. "They end up going into less competitive fields like human resources or marketing, and they don't go for finance, investment banks, or senior-track faculty positions," Brescoll told us. And, as is the case with so many of our female experts, Brescoll used to suffer from the same syndrome herself—until she learned better.

"I've always had to make extra sure I was really, really good," she admitted. "I felt I didn't stack up unless I had more articles in prestigious publications than my male colleagues. But at the same time I would automatically assume that my work wouldn't be good enough for a top publication, that I should aim just a bit lower."

And the men?

"I think that's really interesting," she says with a laugh, "because the men go into everything just assuming that they're awesome and thinking, 'Who wouldn't want me?'"

What Are Men Actually Thinking?

Pretty much that they are awesome, and "who wouldn't want me?" Brescoll is right. Most of the men we interviewed, in addition to our colleagues and friends, say they simply spend less time thinking about the possible consequences of failure.

David Rodriguez is the vice president of human resources at Marriott. For years, he's been our go-to management guru. David has to do a lot of public speaking in his job and loves it. When he takes the stage, his dimples soften his corporate uniform (dark suit, power tie). He says he becomes numb to any criticism from the room. He isn't questioning whether his content is good enough or whether he's flubbed a line or two. He tells himself he's going to ace the presentation, be witty, and impress his bosses. "I just get up there and perform," he said. "The trick is not to overthink it." And if things do go wrong, he shrugs them off. "I don't dwell on stuff; when it's done, it's done." We heard the same attitude from most of the men we talked with. Even when they aren't natural performers, they just move through their challenges with less baggage.

Do men doubt themselves sometimes? Of course. But they don't examine those doubts in such excruciating detail, and they certainly don't let those doubts stop them as often as women do.

If anything, men tilt toward overconfidence. We were surprised to learn that most of the time they arrive at that state quite honestly. They aren't *consciously trying* to fool anyone. Columbia Business School even has a term for it now. They call it "honest overconfidence" and they have found that men on average rate their performance to be 30 percent better than it is.

When we raised the notion of a confidence gap with a number of male executives who supervise women, what we heard was enormous frustration. They believe that a lack of confidence is fundamentally holding women back, but they're worried about saying anything, terrified of sounding sexist. Most don't experience our lack of assurance, they don't understand it, and they don't know how to talk about it.

One male senior law partner told us the story of a young female law associate who was excellent in every respect, except that she didn't speak up in client meetings. His takeaway was that she wasn't confident enough to handle the account. But he didn't know how to raise it as an issue without seeming offensive. He eventually came to the conclusion that confidence should be a formal part of the performance review process because it is such an important aspect of doing business.

David Rodriguez agrees that confidence, expressed or not expressed, even in the most subtle ways, can make or break a rise up the ladder. Among the very top corporate women he deals with, it's not an obvious lack of confidence he sees, because the senior executive women in his organization are quite sure about their abilities. But there's sometimes something he calls a "hesitation." "There's a higher likelihood the women will hesitate at key moments," he suggested. "I think because they often aren't sure what scorecard will be used to judge behavior. And they are afraid to get it wrong. I'll ask later—'what happened at that point in the presentation?' It seemed as though there was a fork in the road. They'll say 'I couldn't get a feel for the audience—how they were responding. I couldn't decide whether to go right or left.'"

A hesitation. It's a fear of failure, perhaps. Or a desire to do it perfectly. Perhaps it's the result of habits formed over decades as the top student. It's also a sense, usually accurate, says Rodriguez, that women *are* being judged by a confusing and shifting yardstick. Or it could be the female brain at work, carefully assessing the emotion of the room. Whatever the causes, that hesitation has consequences. Rodriguez says it can affect whose ideas are adopted, or even who gets a promotion.

There is something so burdensome about the freight of being female that when asked to simply name our gender before a math test, we automatically perform worse. We were floored by the results of one experiment in particular. To explore the impact of "stereotype threat," as it is known, Harvard University gave a group of forty-six highly gifted Asian-American female undergraduates one of three questionnaires, each calibrated to play into different stereotypes. One questionnaire emphasized the stereotype that Asians are good at math, the second emphasized the stereotype that women are bad at math, and the third questionnaire, one administered in the control group, was neutral, emphasizing neither stereotype. After completing the questionnaire, all the women took a difficult math test. The women who were reminded of their Asian heritage correctly answered 54 percent of the questions. Those in the control group answered, on average, 49 percent correctly. The women who were reminded only of their gender scored the lowest, 43 percent right.

We don't really need to read a few paragraphs about women being bad at math in order to stereotype and handicap ourselves in more consequential ways. Hewlett-Packard conducted a study to figure out how to get more women into top management. These numbers say it all: The authors found that the women working at H-P applied for promotions only when they believed they met 100 percent of the qualifications necessary for the job. The men were happy to apply when they thought they could meet 60 percent of the job requirements. So, essentially, women feel confident only when we are perfect. Or practically perfect.

Underqualified and underprepared men don't think twice about leaning in. Overqualified and overprepared, too many women still hold back. And the confidence gap is an additional lens through which to consider *why it is* women don't lean in. Even when we are prepared to tolerate the personal disruption that comes with aiming high, even when we have plenty of ambition, we fundamentally doubt ourselves.

An Even Dirtier Secret

These dark spots of doubt we hide, and sometimes even nurture, need to be vanquished. Confidence is no longer the sideshow, it's the main event.

We might like to believe that keeping our noses to the grindstone, focusing on every detail, and doing everything perfectly are the materials that build a career. Or that overconfidence will lead to ruinous results. In fact, more often than not, the opposite is true.

Cameron Anderson, a psychologist who works in the business school at the University of California, Berkeley, has made a career of studying overconfidence. In 2009, he decided to conduct tests to compare the relative value of confidence and competence.

He came up with a novel idea. He gave a group of 242 students a list of historical names and events, and asked them to tick off the ones they knew. Among the names were some well-disguised fakes: a Queen Shaddock made an appearance, as did a Mr. Galileo Lovano, and an event dubbed Murphy's Last Ride. Anderson found a link between the number of fakes a student picked and how excessively confident the student was. (The very fact that these students picked fakes instead of simply not checking the answers showed they not only were less able, but also believed they knew more than they actually did.) At the end of the semester, Anderson conducted a survey of the group. The students who had picked the most fakes had achieved the highest social status, which Anderson defines as the respect, prominence, and influence an individual enjoys in the eyes of others. Translated into the work environment, he says, higher status means you are more admired, listened to, and have more sway over your organization's discussions and decisions. So despite being the less competent students, they ended up being the most respected and had the most influence with their peers.

His findings upend most assumptions and in some ways they are dismaying. Confidence matters more than competence? We didn't want to believe it, and we pressed Anderson for alternative theories. But deep down we knew we'd seen the same phenomenon for years. Within any particular organization, from the boardroom to the PTA, some individuals tend to be more admired and more listened to than others. They're the ones at meetings who lead the discussion and often dictate the outcome. Their ideas kick up to the next level. They are not necessarily the most competent people in the room; they are just the most confident.

More disturbing for women who count on competence as the key to success is Anderson's insistence that actual ability barely matters. "When people are confident, when they think they are good at something, regardless of how good they actually are, they display a lot of nonverbal and verbal behavior," Anderson explained. He mentioned their expansive body language, their lower vocal tone, and a tendency to speak early and often in a calm, relaxed manner. "They do a lot of things that make them look very confident in the eyes of others," he added. "Whether they are good or not is kind of irrelevant."

It's confidence that sways people. We may not realize it but we all give confidence inordinate weight and we respect people who project it. Anderson is convinced this explains why less competent people are so often promoted over their more able colleagues. Infuriatingly, there aren't even necessarily negative consequences for that lack of competence. (This all explains a lot about high school.) Among Anderson's students, confidence without competence had no negative effects. They were simply admired by the rest of the group and awarded a high social status. "The most confident people were just considered the most beloved in the group," he said. "Their overconfidence did not come across as narcissistic."

That is a critical point. Overconfidence can also be read as arrogance or bluster, but Anderson thinks the reason the more confident students didn't alienate the others is that they, like those Columbia Business School students, *weren't faking their confidence.* They genuinely believed they were good, and that self-belief was what came across. Fake confidence, he told us, just doesn't work in the same way because we can see the "tells." No matter how much bravado they muster, when people don't genuinely believe they are good, we pick up on the shifting eyes and rising voice and other giveaways. We're not always conscious of it, but most of us have a great BS radar and can spot fake confidence a mile off.

We both wondered, perhaps a little vindictively, whether

Anderson believes that overconfident people are just stupid. Are they at all aware that their confidence outstrips their abilities? They may, in fact, be less intelligent, he conceded, but he also pointed out that he's focusing on a relatively modest amount of overconfidence. Even a popular pilot has to be able to land a plane. If the gap between confidence and competence grows too large, overconfidence does become a weakness and a liability. But that's not a problem most women need to worry about.

Once we got over our feeling that Anderson's work suggests a world that is deeply unfair, we could see a useful lesson: For decades, women have misunderstood an important law of the professional jungle. Having talent isn't merely about being competent; confidence is actually a part of that talent. You have to have it to be good at your job.

Getting to the Zen of It

When we aren't confident, we don't succeed as we should. We can't even envision the work we could be doing, or the levels we could reach, or the satisfaction we could have. We can't contribute fully to a system that is in great need of female leadership.

But confidence provides so much more than that. It tends to get unfairly tagged as a showy quality that is all about competition and outward success. We found it has a much broader impact. Scholars are coming to see it as an essential element of internal well-being and happiness, a necessity for a fulfilled life. Without it you can't achieve flow, the almost euphoric state described by psychologist Mihaly Csikszentmihalyi as perfect concentration; the alignment of one's skills with the task at hand. Flow is like being in the athletic zone; it is a state of mastery impossible to reach without confidence.

A Buddhist meditation center made for a welcome moment of calm after the noisy and unsettling explorations of sports arenas, military schools, and collegiate lecture halls. We were looking for insight into what confidence might do for us as people, as social animals, beyond winning games and points with the boss, and we hoped we might find it in a more spiritual realm.

Sharon Salzberg, a leading Buddhist expert, author of a number of best sellers, including *Real Happiness at Work*, and a friend of a friend, was in town leading a meditation session. We found her on the top floor of a five-floor walk-up, in a wood-paneled room full of warmth and light, holding forth on the virtues of equanimity. Three dozen extremely centered-looking students filled the room and we were quickly inspired to put all hard questions aside for an hour, and go limp.

When Salzberg later turned to our subject, we heard something that clicked. "I think confidence is the way we meet our circumstances, whether they are wondrous and wonderful or really hard and difficult," she offered, with a tranquil smile. "It's almost like a wholeheartedness, where we're not holding back. We're not fragmented. We're not divided. We're just going towards what's happening. There's an energy to it. I think that's confidence. And it's absolutely part of human fulfillment." We were captivated by the idea of confidence as an essential, elemental energy. It was a twist, but it fit. We realized we were beginning to amass some related, yet different descriptions—"purity of action," "wholeheartedness," "energy." They seemed to be pushing us toward a very basic question: What exactly is confidence anyway? Before we spent any more time trying to catch it in action, examine where it comes from, or ask why women have so little, we figured we'd better give confidence a formal definition.

2

DO MORE, THINK LESS

Like us, neuroscientist Adam Kepecs is searching for confidence. But, unlike us, he has a preference for small, furry rodents. Rats, says Kepecs, are less complicated than people. They don't bury their basic instincts in layers of tangled thought and emotion. People will tell you they are confident, when, inside, they're quivering wrecks. Or the opposite. They'll tell you they feel insecure, but then their actions suggest boldness. As research subjects, Kepecs finds people unsatisfactory.

He is trying to get to a notion of confidence that is very basic: He calls it "statistical confidence," or, in layman's terms, the measure of our certainty about a choice we've made. His groundbreaking studies have caught the attention of psychologists because they suggest confidence is a quality all species possess. Who knew rats could be confident, too?

We were intrigued by Kepecs's work, and hoped that what he

explores in rats might help us understand what constitutes basic confidence in humans. Confident decision making in rats, he believes, shares many similarities with human decision making.

Imagine, he told us, that you're driving to a new restaurant. You've been given the directions and, at the light, you make a turn. You drive a mile, and then another mile. No restaurant. At some point you start thinking, "I'm sure I should be there by now. Did I make a wrong turn?" Whether or not you stick it out and keep driving depends on how confident you are of that turn you made. It's that "sticking it out" piece that Kepecs measures in the rats' behavior, and it suggests that confidence, stripped down, is a pretty basic commodity.

What is confidence, really? Well, it's certainly not what we anticipated it was when we started researching this book.

Confidence is not, as we once believed, simply feeling good about yourself, saying you're great, perfect just as you are, and can do whatever you want to do. That way of thinking hasn't really worked for us, has it? Just saying "I can do that" doesn't mean that you believe it or will act on it. If it did, therapists would be out of business pretty quickly. And hearing "You are wonderful" from someone else doesn't help, either. If all we needed were a few words of reassurance, or a pat on the back, we'd all be productive, thin, and nice to our in-laws as we commandeered the corner office.

We also had a vision of confidence as a set of mannerisms and an expression of power. The most confident person seems to be the one who speaks the loudest and the most often. The friend who always knows he is right or the colleague who dominates every meeting. Aren't those the most confident people, the ones who just, well, sound so confident?

We were counting on Kepecs to help us out, and we met him

at Cold Spring Harbor Laboratory, a dazzling setting, right on the ocean on Long Island, forty-five minutes east of Manhattan.

Dangling over our heads as we wound our way to his upstairs office was an enormous, Dale Chihuly glass sculpture—a gift to DNA pioneer James Watson, who transformed the research center into one of the best in the world. The swirl of luminous yellow and green tentacles, capped with bubbles of varying sizes and shapes, demanded a moment of contemplation. Extremely Dr. Seuss, we remarked. Kepecs, a blue-jeans clad, boyish thirty-nine, with dark curls and a hint of Hungarian in his voice, laughingly explained that the sculpture is actually an ode to the shape of neurons. Naturally.

For the next few hours, he was our generous translator of this unfamiliar world—helping us look for connections between his rodents and the human confidence code.

We watched as Kepecs put a rat in a large box. The rat was wearing what is essentially a permanent white top hat housing an array of electrodes. It had been surgically attached, and Kepecs assured us that the rats no longer feel it at all. Inserted into one side of the box, level with the rat's nose, were three white containers, or ports, about two inches wide. The middle one released odors. The rat put its nose into that port and sniffed a mix of two smells. The mix varied in percentages; sometimes the stronger smell was clear and, at other times, teasing apart the combination was trickier. The rat's job was to figure out the predominant smell, and then put his nose in either the left or right port to indicate his decision. If he gets it right, Kepecs explained, and chooses the correct port, he'll get a drop of water as a reward. But the rat has to wait for the drop of water. If he's sure about his decision, he'll wait as long as it takes for the water to come. If he's doubtful, he can give up on getting that particular drop, and start a new round. But giving up means the rat loses not only the chance of getting the drop, but also all of the time it already invested in waiting for the drop. The rat faces a real trade-off, a fundamental, familiar dilemma, and one that turns out to be shared across species. We watched our rat put his nose to the left, and then wait for what seemed an endless . . . eight seconds. That's a long wait for a rat, and so there was plenty of confidence on display. Would it prove to be justified?

Bravo! We exchanged a smile as the drop of water materialized. Kepecs warned us not to get any ideas about how "smart" the furry creatures were. The rats in this particular experiment had done the drill countless times, and they were all pretty good at figuring out which odor corresponded to the left or right port. Kepecs, remember, isn't focused on *whether* his rats make the right choice. He's measuring *how firmly they believe* they've made the right choice. That is the confidence Kepecs works to isolate—the strength of a rat's belief in its decision. It's a confidence that is demonstrated by a rat's act of waiting, and then measured by the length of time it's willing to stick it out, braving a real risk of failure, waiting for that drop of water. It was extraordinary to us that not only could these rodents apparently make what was a calculation about odds and stakes, but that they were then willing, essentially, to bet on their decision.

There is something elemental to this expression of confidence. The rats are making an informed prediction, almost robotic in its execution. Human brains, too, at times, can also act almost robotically. Every day we make hundreds of decisions, almost unconsciously, that require basic confidence—how quickly to reach out to hit the snooze button on our alarm clock, how far to bend over to load the dishwasher. Kepecs has pinpointed the part of the brain that rats use for these decisions, the orbitofrontal cortex, and he thinks statistical confidence for humans will be found to hail from the same region.

What we saw in Kepecs's lab at Cold Spring Harbor sharpened our picture of confidence. For one thing, if the rats were to be believed, it was not merely a brand of aggressive behavior, or an endless focus on feeling good about oneself. A rat's confidence might be broadly described as a belief that it can create a successful outcome (drops of water) through its action (waiting). We saw a hint of self-efficacy in that. It's a chain of events that all starts, we observed, with that very basic, perhaps unconscious, confidence calculation, which then encourages the rest of the action.

Kepecs gave us a deeper take on confidence, for both rats and humans. In his view, confidence has a distinctive double nature, or shows "two faces." One face is objective: that basic calculation process, a critical confidence tool we'd watched the rats employ. The other face, Kepecs told us, is subjective. Confidence is also something we experience as a feeling. *That's* the confidence we're more familiar with, and spend a lot more time around, at least consciously. It's the more emotional element, its alluring promise yet illusory nature constantly tripping us up. Rats too, Kepecs believes, *feel* their confidence in some ways.

It occurred to us, in the middle of this engrossing and enlightening session, that women would be well served by spending a bit more time rubbing shoulders with Kepecs's version, or really any of confidence's other lesser-known, less glamorous, more workman-like renderings. Maybe confidence shouldn't be so mysterious and glamorous and perversely aspirational-only. How refreshing to view it, at least in part, as a simple, concrete tool: an extremely useful compass, perhaps, if we could just get the darn thing working.

Naturally, we started wondering how we measured up. Were we as confident as the rats, at least? We asked the ever-patient Dr. Kepecs whether he could measure our basic, objective confidence somehow, without surgically implanting electrodes or forcing us to inhale a lot of questionable odors. Kepecs had been running some similarly themed tests on students using only computer games. That sounded good to us. As we attempted the various, unfamiliar screen challenges though, we were each surprised to feel considerable anxiety about how we were performing. We found out pretty quickly that we'd both scored extremely well—both in our statistical confidence (measured by how long we took to rate our confidence in each answer) and in our actual accuracy. But just before we got that news, the two of us compared notes, confessed our nerves, and predicted to each other that we'd both bombed. And, at that moment, we really meant it. Sigh.

It was our own twisted version of that familiar paradox. We found our participation in it hard to believe. We'd read so much research about women doubting and underestimating their performance on tests, but we still couldn't fully avoid experiencing it. There we had been, performing perfectly well, not only answering questions fairly accurately, but also simultaneously reporting high levels of confidence in those answers, and yet we still experienced palpable self-doubt, and *told ourselves*, and each other, that we expected we'd done poorly. What is that? Maybe, we speculated, the female subjective and objective confidence wiring is just totally crossed somehow. We also wondered whether this disturbing pattern of behavior is confined to the human female.

We were starting to stray into the psychological, the philosophical and the bizarre with a host of human-specific questions when Kepecs reminded us that the rat/human comparison is limited.

Confidence is clearly more labyrinthine for higher-order, abstract thinkers. Rats don't brood, for example, second-guess themselves, or lie in bed frozen with indecision. And they don't suffer from a gender-driven confidence gap either. There was only a certain amount Kepecs, and his influential work, could explain for us. It was time to look at confidence outside the lab.

Into the Wild

On a day when the rest of the Georgetown student body was outdoors, enjoying an unseasonably warm spring sun, we found a dozen young women cooped up in a classroom, learning how to run a political campaign. They were there thanks to a nonprofit organization called Running Start, which was founded to train college-age women to run for public office. The women were smartly dressed and not timid exactly, but quiet and serious. When we arrived, they were huddled in groups of three or four discussing their motivations for running for college positions.

One girl was upset that condoms weren't sold on campus; another that there were no rape kits. One student worried about how the college endowment was used; another about how research positions were allocated. A Running Start facilitator, Katie Shorey, led the conversation, gently guiding the separate discussions: "If you were to run, what would you talk about and try to change? How passionately do you feel about this issue?"

These were women who wanted to change the world, and they were nurturing aspirations of running for political office. They were among the best and the brightest, or they wouldn't be at Georgetown. We joined the class that day expecting to meet some of the country's most self-assured young women, hoping they'd help us define confidence.

What struck us immediately was how polite and considerate they were. They didn't just jump into the discussion; they raised their hands first, asking, "May I add something?" or "Can I suggest this?" We couldn't help thinking how different it would be with a group of men. Would they ask permission before speaking? Most young men would be louder, more assertive, keener to make sure their opinions were heard. Men's attention to good manners might be lacking, and their rude interruptions annoying (at least to women), but their conversations, we suspected, might be less cautious. Not for the first time, we wondered about the tipping point between assertiveness and jerkiness. To put it bluntly—does one have to be an asshole to be confident?

As the women gathered back into one group, we put a question to this room of diligent, high-scoring achievers: Who among them feels confident about running for a post on the student body? Not a single hand went up. So, we asked: What is it that makes you feel so nervous? And, in their answers, these Georgetown students painted a vivid picture of all the things confidence is *not*.

> "Running for office means we have to self-aggrandize. That's hard because people might think we're pushy."

> "If I lose, it'll be about me, because they don't like me."

> "I internalize setbacks. The other day a professor criticized my research paper. The guy I'd worked on it with just brushed it off. It didn't seem to bother him. It took me weeks to get over it."

"I ran for a position in high school once with a guy student and we won. I was more shy and he was more confident, but I did all the work. The next year we ran against each other, and I lost. But I know I was the more competent one. I did all the hard work. It was a real blow."

"If a woman is assertive and ambitious, she's seen as a bitch. But for a guy, hey, those are normal qualities."

"I went to an all-girls school. It was so empowering because everyone who raised their hand in class to ask a question was a girl. It was normal. But then, I came here; I saw that the girls didn't speak up in class. And here's what's really sad—started copying them. I started raising my hand less and self-censoring—just in order to fit in."

After the uncomplicated rats (who show no obvious gender bias in confident decision making, by the way), this conversation was a letdown. We recognized, yet again, what a waste of energy and talent all of this agonizing can be. We discussed it with Jessica Grounds, the cofounder of Running Start, who recently joined the powerhouse Ready For Hillary political action committee to handle all of their outreach for women. She told us that her team had come to realize, over the years, that what these ambitious young women need most isn't a primer in running for office as much as basic confidence training. They have the skills; what they lack is self-belief, and without it they can't turn their *desire* to run into the *action* of running. If they don't take the chance, they will be stuck, spinning around inside their heads like Adam Kepecs's rats on a treadmill.

We say this not with contempt, but with the recognition that many of their anxieties rang true to us personally.

We both spent too much of our twenties and thirties stuck in self-doubt, and yes, we both still devote too much time to internalizing setbacks. After delivering a speech not long ago to applause and compliments, Claire spent a good hour wondering why two women, in a room of more than a hundred, had looked somewhat bored. For the sake of their own sanity and happiness, young women have to find a way to interrupt that negative soundtrack—much sooner, we hope, than we've been able to.

Five-Star Confidence

A host of bureaucracy and formality comes with a visit to one of the most senior women in the U.S. military: multiple security checks, multiple escorts through a matrix of endless corridors, thickly paneled with rousing paintings of seminal battles and imposing portraits of heavily decorated, square-jawed, almost exclusively male, generals and admirals. And then there's that mouthful of a title: The Undersecretary of Defense for Personnel and Readiness. When we finally made our way to Major General Jessica Wright, however, in her suite of offices tucked deep inside the Pentagon, she was surprisingly, refreshingly, not what we expected. While her office décor is predictably masculine—leather club chairs and mahogany tables—Wright was anything but. She may be top brass, but she was down-to-earth, guiding us to a sofa and putting us at ease with a few questions.

The general's eyes were bright and inquisitive, and she made a point of listening carefully. There was no bluster or overly aggressive assertiveness about her; she doesn't condescend. She was also resolutely feminine, no doubt following one of the top ten leadership tips she shares with other women: Enjoy getting your hair and nails done. Just because you're working in a man's world, she laughed, doesn't mean you always have to look like them. We liked that. Wright wasn't curbing her own personality to try to fit into a mold; she had the moxie to be true to herself. Daring to be bold about girlfriendy things like manicures and blowouts while holding a general's rank in the most powerful military in the history of the world seemed pretty confident to us.

Another thing we really liked about Wright's style of confidence: She was prepared to admit to nerves, but she didn't let them stop her from pursuing her goals and ambitions. She described taking command of an army combat brigade in 1997, the first woman to do so, and feeling so nervous she could barely breathe. "My mother taught me to be stoic," she said with a smile, "but my insides were a spaghetti bowl of feelings and confusion and anxiety."

We should dispel any impression that General Wright is an angst-ridden pushover, though. You don't get to her position without a certain amount of grit. She doesn't hesitate often. She told us that good leadership means being an efficient decision maker, and she doesn't tolerate indecision in others. "When somebody says to me, 'Well, I don't know what to do,' I don't have time for that. Because if I ask you to give me your opinion and you're wishy-washy with me, I'm moving on. We're always on a fast-moving train," she said, crisply, and we got a sense she's not somebody you'd want to let down.

Or underestimate. Because, when pushed, Jessica Wright acts, even when intimidated. She remembers the occasion when she was a brand-new lieutenant, and a superior told her straightaway that he didn't like females in the military. "There were five hundred things going through my head," she said. "And I looked at him and said, 'Sergeant Minski, you have an opportunity now to get over that.'" She smiled mischievously at our laughter. "I still don't know how that came out of my mouth. I really don't."

Her bold retort paid off. The misguided sergeant and she became friends, and he even mentored her as a young officer. She puts it down to that first encounter, when she proved she could, and would, stand up for herself.

We'd paid a visit to Wright hoping she'd help us characterize confidence. In the end, she didn't even have to describe it for us because she demonstrated it so clearly in her style, her stories, and her observations. In our notebooks, in addition to having drawn big bold circles around her Sergeant Minski quote, which we were eager to appropriate, we'd jotted down—*action* and *bold* and *makes decisions*. But we'd also written *honest* and *feminine*. And also this: *comfortable.* General Wright has the layers of emotional complexity we didn't find in Adam Kepecs's lab, but she has overcome the torment of those Georgetown students. To us, she had a handle on what we were starting to understand is real confidence.

Life Lessons on a Kiteboard

We had a nascent theory in the works, and we wanted to try it out on the experts, the psychologists who make this subject their life's work. We started by asking them this seemingly simple question: How do you define confidence? Time and again, we were met with a long pause, followed by, "Well, it's complicated."

"Confidence," said Joyce Ehrlinger, the University of Washington psychologist, sighing in sympathy, "has become a vague, almost stock term that can refer to any number of things. I can see why you'd be confused."

"General confidence is an attitude, a way you approach the world," suggested Caroline Miller, a best-selling author and positive psychology coach. "More specifically, self-confidence is a sense that you can master something."

"One way to think about confidence," said Brenda Major, the UC Santa Barbara social psychologist, "is how sure are you that you have the skills that you need to succeed in doing a particular thing."

"It's a belief that you can accomplish the task you want to accomplish," Utah State University's Christy Glass told us. "It's specific to a domain. I could be a confident public speaker, but not a confident writer, for example."

Glass's observation helped us to understand why confidence can seem like such a fleeting quality. In some circumstances, we have it; in others, we don't. It explains how Andre Agassi, for example, could be so incredibly confident about his tennis but so riddled with self-doubt in the rest of his life. It explains why so many women might feel confident in their personal lives but not at work, and it explains why Claire can be confident in her people skills but not as self-assured when it comes to making decisions. She doesn't overthink when she's helping other people solve problems, but has trouble solving her own.

Caroline Miller's mention of mastery also got our attention. Initially, we were wary, and somewhat suspicious of the term. It sounded undeniably masculine and evoked images of paternalistic gentry lording over their subjects. It also seemed like something we might need power tools for, not to mention a high school shop course. Our real fear, though, was that mastery would just turn out to be a recipe for the endless pursuit of perfection—something to which women are far too susceptible already.

But Miller explained that mastery is none of that. Mastery isn't about being the best tennis player or the best mom. The resonance of mastery is in the *process* and *progress*. It is about work, and learning to develop an appetite for challenge. Mastery inevitably means encountering hurdles; you won't always overcome them, but you won't let them stop you from trying. You may never become a world-class swimmer, but you will learn to swim across the lake. And the unexpected by-product of all of that hard work you put in to mastering things? Confidence. Not only did you learn to do something well, but you got a freebie.

This next point is invaluable. The confidence you get from mastery is contagious. It spreads. It doesn't even really matter what you master: For a child, it can be as simple as tying a shoe. What matters is that mastering one thing gives you the confidence to try something else.

When Katty turned forty, for example, in defiance (or perhaps denial) of middle age, she made the decision to learn to kiteboard. She needed a challenge, and had a naïve fantasy that if she could crack this, she'd soon be a cool (young) surfer chick doing acrobatic jumps high above the waves. She didn't, however, anticipate how often she'd get dragged down a beach attached to a powerful thirty-foot kite, or fall from her board into the saltwater, or the tears and frustration and loud cursing. After the first couple of summers, she was prepared to give up; it was too humiliating and she was too sore. But she stuck with it and, while youth and coolness remain slightly beyond her grasp, she can now kitesurf. Her children, who started long after her, are naturally already ten times better, but that's not the point. Having mastered (sort of) one extreme sport, Katty's now looking around for another to help face down the next decade.

Meet the Confidence Cousins

Confidence was starting to come into focus. We were growing convinced that it involves action—doing, mastering, maybe even deciding, but we still had a jumble of other terms fighting for our attention. (For a while, we even made the rookie mistake of using "confidence" and "self-esteem" interchangeably.) Our experts set us straight. The confidence cousins are all worth having as well, but there are some critical differences between confidence and the other positive attributes that many of us tend to lump together: self-esteem, optimism, self-compassion, and self-efficacy.

Some of these extended family members have been pored over and heavily researched. Others are new to the scene. They each have their detractors and supporters. Some people will tell you that optimism is the key to life; others are equally adamant that, without self-esteem, you will never be happy. What they do have in common is that each allows us to improve the richness of our life, to function at maximum capacity, to enhance our professional performance, and to deepen our personal relationships. In an ideal world we'd all have all of them in abundance.

Self-Esteem

"I am a valuable person and I feel good about myself." Agree with that and chances are you have pretty high self-esteem. This is a value judgment on your overall character. It's an attitude: "I like myself," or "I hate myself," or more typically something between the two extremes. Self-esteem is what allows us to believe that we are lovable, that we have value as human beings. It's not related to wealth: You can be the richest, most successful CEO in your industry and have low self-worth, and you can be a cashier at a drugstore and have plenty of self-esteem.

In the mid-1960s, sociologist Morris Rosenberg came up with a basic self-esteem scale that is still the worldwide standard. It's a simple list of questions: "I feel I do not have much to be proud of," "I feel that I have a number of good qualities." Answer those, and eight others, and you can quickly measure your self-worth. We've put the scale in the notes, if you are curious about how you stack up.

Self-esteem is essential for emotional well-being, but it is distinct from confidence because confidence is typically tied to feelings about what we can achieve: "I am confident that I can run this race and get to the end." Self-esteem tends to be more stable and more pervasive than confidence. If you have an overall good feeling about your position in the universe, chances are you'll have that for life and it will color much of what you do. It's an invaluable buffer for withstanding setbacks.

There is considerable overlap with confidence, to be sure. A person with high self-esteem will tend to have confidence, and vice versa. There's a particularly close relationship if high self-esteem is based on talents or abilities. "I think I'm a valuable person because I am smart, fast, efficient, and successful in my field." If, however, you don't really care about talents or skills or intelligence or achievements but you care about being a good person, perhaps about being devout, or living up to a moral code, then your self-esteem and your confidence will have looser ties.

It is worth noting that there's been a self-esteem backlash lately; the concept has developed a bad name in the minds of psychologists (not to mention employers, teachers, parents, and, even some of its former promoters) after a decades-long push into schools, homes, and even workplaces. That's because the product being pushed was unrealistic self-esteem. The emphasis was on simply telling children, and sometimes adults, they were all winners, all fabulous, and all perfect. After observing a generation of self-esteem-swaddled kids turn into rudderless adults, the experts realized none of that actually gives children any concrete basis for believing they can do anything, or even make decisions on their own.

Optimism

Optimism has muscled self-esteem aside and is the hotter commodity these days. In Latin, the word optimum means most favorable. So an optimistic person is one who expects the most favorable outcome from any given situation. Optimism is a question of interpretation, and that basic glass half-full, half-empty measure still works well. We can experience the same facts—the glass has the same amount of water—but how we interpret those facts depends on our optimistic or pessimistic attitude. Winston Churchill put the difference memorably: "A pessimist sees the difficulty in every opportunity; an optimist sees the opportunity in every difficulty." Another hallmark of optimism is gratitude. If you are an optimist, you notice that good things happen to you, and you feel grateful for them. If you're a pessimist, you probably don't pay attention to positive things as often, and when they do happen, you believe them to be chance occurrences. Psychologists suggest this simple test: Open a door for an optimist, and the chances are she will thank you. A pessimist is much less likely to even notice the door being held for her and, if she does, she will assume it was merely being opened for someone else.

You can be optimistic about a specific event: The marathon will be fun, or the test will be easy for me. Or you could have a general view that things will work out positively. Unlike self-esteem, optimism isn't a judgment on your inner self-worth; it's an attitude you have that is based on your view of the outside world. You're not optimistic because of your talents or your innate goodness; you are optimistic because you interpret the world positively.

Nansook Park is one of the world's leading experts on optimism and a professor at the University of Michigan. She describes confidence and optimism as closely related but with an important distinction—optimism is a more generalized outlook, and it doesn't necessarily encourage action. Confidence does. "Optimism is the sense that everything will work out," she says. "Confidence is, '*I* can make this thing work.'"

One of the most influential voices in psychology today is that of Martin Seligman, one of the founders of the positive psychology movement. He's redefined optimism as something more robust, with a sense of action, tugging it closer to confidence. In his best-selling book *Learned Optimism*, he contends that optimism, like other skills, can be cultivated through, among other things, mastery and hard knocks, which help to develop a sense of personal agency. Optimistic people, in Seligman's view, have a sense that they can effect change. Therefore the world does not appear as bleak.

Self-Compassion

Self-compassion is the newest, edgiest member of the extended confidence family and, at first introduction, can seem reminiscent of the groovy, hippie 1960s. The concept springs from the Buddhist theory of loving-kindness and the work of Sharon Salzberg, but was recently pioneered as an academic pursuit by Kristin Neff, a professor in the educational psychology department at the University of Texas. The central precept is that we should all be kinder to ourselves because doing so makes us healthier, more fulfilled, and more successful in the pursuits we choose.

Self-compassion dictates that we treat ourselves as we treat our friends. If your friend comes to you and says, "I just failed. I blew it," what do you do? You're kind, you're supportive, you're understanding, and you give your friend a hug. Or if it's a guy, you give your friend a slap on the back. You try to pick the other person up. But, Neff told us, all too often we don't do that for ourselves: "Indeed, often the people who are most compassionate toward others are the least forgiving to themselves."

The second key to self-compassion is that it places our individual experiences in the framework of a shared human experience. It takes our imperfections and sufferings and puts them in the context of simply being human. In our success-oriented world, we tend to think of failure as abnormal. We get a low grade, get turned down for a promotion, lose our job, or get dumped by our boyfriend, and our instinct is to say, "This shouldn't be happening." But, of course, these setbacks are just part of being human and if you never had them, you'd be a robot. Putting our disappointments in that context makes them less frightening and less isolating.

So how does self-compassion fit into the confidence clan? At first glance, self-compassion and confidence seemed an ill-suited pair. Confidence, we were now fairly certain, involves action. Self-compassion says, "Don't beat yourself up; put yourself in the broader human condition, and accept some failure." We wondered why self-compassion doesn't just encourage us to accept all our flaws so thoroughly that we become comatose. Why not just stay on the couch and surf the shopping channels? "I was mean to my friend, failed to make dinner, didn't talk to that lonely-looking person, didn't do my homework, didn't go to the gym, didn't finish my project at work, didn't take the tough college course. Oh well, I'm only human. Where's the remote?"

Neff patiently explained that far from being in conflict with confidence, or encouraging sloth-like behavior, self-compassion drives confidence—allowing us to take the very risks that build it. It is a safety net that actually enables us to try for more and even harder things. It increases motivation because it cushions failure.

"Most people believe that they need to criticize themselves in order to find motivation to reach their goals. In fact, when you constantly criticize yourself, you become depressed, and depression is not a motivational mind-set," Neff said.

Having overcome our initial, overachiever reservations, there's something else appealing about self-compassion. It is the acceptance that it's okay to be average sometimes. Many of us spend our lives trying to be the best at everything, whether it's winning soccer games at age five or making partner by age thirty-five. We live in a culture where being anything other than the winner is frowned upon.

"If I were to tell you that your work as a journalist is average, you'd be devastated, right?" Neff said. "Being called average is considered an insult. We all have to be above average. It sets up a very comparative mind-set. But the math doesn't work. It isn't possible for us all to be above average, even though many studies show most Americans think they are."

We live in a world of constant comparisons that extend well beyond the workplace. She's thinner, richer, and more successful than we are. She's a better mom, has a better marriage. But constantly defining yourself through other people's achievements is chasing fool's gold. There is always someone doing it better. Sometimes you fare well by comparison; sometimes not.

Self-compassion recognizes the folly of this. To take risks, we have to know that we won't always win. Otherwise, we'll either refuse to act or be devastated. Self-compassion isn't an excuse for inaction—it supports action, and it connects us to other people, to being human, with all the strengths and the weaknesses that implies.

Self-Efficacy

If self-compassion is the kind, gentle cousin, self-efficacy is the tough, just-get-it-done member of the family.

In 1977, psychologist Albert Bandura's article "Self-Efficacy: Toward a Unifying Theory of Behavioral Change" was published. In the sedate world of academic psychology, this work, with its arcane title, sent tremors throughout the field. For the next thirty years, self-efficacy was one of the most studied topics in psychology. *Self-efficacy* is defined as a belief in your ability to succeed at something. Bandura's central premise was that those beliefs, our sense of self-efficacy, can change the broader way we think, behave, and feel. Self-efficacy, much like mastery, creates spillover effects.

Self-efficacy's goal-oriented nature especially appealed to the success-focused baby boomer generation. But it's also a simple and practical quality. We can all identify specific goals we want to achieve: lose twenty pounds, learn Spanish, and get a pay raise. Bandura says the key to actually putting those aspirations into action is self-efficacy.

If you have a strong sense of self-efficacy, you will look at challenges as tasks to be conquered; you will be more deeply involved in the activities you take on, and you will recover faster from setbacks. A lack of self-efficacy leads us to avoid challenges, to believe that difficult things are beyond our capability, and to dwell on negative results. As is the case with confidence, mastery is fundamental to self-efficacy. In other words, try hard, become good at something, and develop self-efficacy—a belief that you can succeed.

Some experts told us they see self-efficacy as interchangeable with confidence. Others maintained that there are distinctions, that confidence can also be a much more generalized belief about your ability to succeed in the world. Self-efficacy also sounded, to us, a bit like Seligman's view of learned optimism. All three are closely tied to a sense of personal power.

Whatever formal label you put on it, whether it's a slice of self-efficacy or a component of optimism or an element of classically defined confidence, that belief that you can succeed at something, that you can make something happen, resonated right away with us. It fit with our observations about action. It seemed to be a central strand of the confidence we were after.

The Real Thing

You know that old saying, "It's all in your head"? Well, when it comes to confidence, it's wrong. One of the most unexpected and vital conclusions we reached is that confidence isn't even close to all in your head. Indeed, you have to get *out of your head* to create it and to use it. Confidence occurs when the insidious self-perception that you aren't able is trumped by the stark reality of your achievements.

Katty discovered this reality in a high-octane, underventilated

White House back office. She was called to attend a briefing of Middle East experts at which she *felt* like the only unqualified person in the room. "The high-powered setting made me insecure," she admits. "When it came to question time, I wanted to ask something, but was worried I'd sound uninformed, that I might blush or seem stupid." It was mostly men in the room and they all sounded so sure of themselves. The easiest path was to do nothing and keep quiet; the risky thing, the confident thing to do, was to speak up. Eventually, after a ridiculous amount of internal agonizing, she did get a question out. "I realized I just had to physically force my hand up, keep it there, and get the words out. And guess what, the sky didn't fall on my head! My question was just as smart as anyone else's. Now, whenever I'm in that position I tell myself I *did* it once, I can *do* it again. And every time, it gets a little bit easier."

We'd seen it in the rats, and we heard it from General Jessica Wright and the academics: Confidence is linked to doing. We were convinced that one of the essential ingredients in confidence is action, that belief that we can succeed at things, or make them happen. Confidence, we saw from the young women in Running Start at Georgetown, is not letting your doubts consume you. It is a willingness to go out of your comfort zone and do hard things. We were also sure that confidence must be about hard work. Mastery. About having resilience and not giving up. The confidence cousins can all support that goal. It's easier to keep going if you are optimistic about the outcome. If you have self-efficacy in one area, and use it, you will create more general confidence. If you have high self-esteem, and believe you are intrinsically valuable, you won't assume your boss thinks you're not worthy of a raise. And, if you fail, self-compassion will give you the chance not to berate yourself, but to take your failure more lightly.

We were at last confident about the way we wanted to define confidence. We felt all the more so when one of our most stalwart guides through this tricky terrain, Richard Petty, a psychology professor at Ohio State University, who has spent decades focused on the subject, managed to put all we had learned into appealingly clear terms: "Confidence is the stuff that turns thoughts into action."

Other factors, he explained, will of course play a role. "If the action involves something scary, then what we call *courage* might also be needed for the action to occur," Petty explained. "Or if it's difficult, a strong will to persist might also be needed. Anger, intelligence, creativity can play a role." But confidence, he told us, is the most important factor. It first turns our thoughts into judgments about what we are capable of, and it then transforms those judgments into actions.

Confidence is the stuff that turns thoughts into action. The simplicity was gorgeous and compelling. It immediately became not only our definition, but an organizing principle for the next phase of our exploration. And what was especially useful was that it somehow, naturally, effortlessly, made proper sense of the other threads we'd been gathering. The critical link between confidence and work and mastery suddenly made sense. They form points on a wonderfully virtuous circle. If confidence is a belief in your success, which then stimulates action, you will create more confidence when you take that action. And so on and so forth. It keeps accumulating, through hard work, through success, and even through failure.

Maybe Nike has it right. At some point we have to stop thinking, and just do it.

We found a striking illustration of how this might play out in the real world (or in something edging closer to the real world) in Italy, at the University of Milan. There we tracked down psychologist

Zach Estes, who's long been curious about the confidence disparity between men and women.

A few years ago, Estes did a series of tests that involved getting five hundred students to reorganize a 3-D image on a computer screen. It looked like a simplified Rubik's cube. He was testing a few things—the idea that confidence can be manipulated and that, in some areas, women have less of it than men.

When Estes had the students, men and women, solve a series of these spatial puzzles, he found that the women scored measurably worse than the men. But when he looked back at their actual answers, he found the reason the women were doing less well was that they didn't even attempt to answer a lot of the questions. They simply ducked out because they weren't confident in their abilities. He then told them they had to at least *try* to solve all the puzzles. And, guess what: The women's scores shot up, and they did as well as the men. Crazy. Maddening. Yet also hopeful.

Estes's work illustrates, in a broad sense, an interesting point: The natural result of under-confidence is inaction. When women don't act, when we hesitate because we aren't sure, even by skipping a few questions, we hold ourselves back. It matters. But when we do act, even when we're forced to act, to answer those questions, we do just as well as men. The women in Estes's experiment skipped questions because they didn't want to try something at which they thought they might fail. In truth, they had no need to worry. They were just as good at manipulating those computer images as the men. But fear of failure led to inaction, thus guaranteeing failure.

Using a different test, Estes simply asked everyone to answer every question. Both men and women got 80 percent right, suggesting identical ability. He then tested them again and asked them, after each question, to report their confidence in their answer. Just having to think about whether they felt certain of their answer changed their ability to do well. Women's scores dipped to 75, while the men's *shot up* to 93! Are women really that susceptible to seizing any chance to think badly of themselves? One little nudge asking us how sure we are about something rattles our world, while with men, it seems to just remind them that they're terrific.

Finally, Estes decided to attempt a direct confidence boost. He told some members of the group, completely at random, that they had done very well on the previous test. On the next test they took, those men and women improved their scores dramatically. It was a clear measure of what confidence can do—fuel our action, and substantially affect our performance, for better or for worse. And we can all imagine, without much trouble, what this suggests about women and confidence in our everyday lives.

Life's Enabler

Think about it. We are all capable of imagining how great it would be to write that novel, apply for that new position, or just introduce ourselves to that interesting stranger. But how many of us actually do it?

Confidence is life's enabler—professionally, intellectually, athletically, socially, and even amorously. The man you met at a conference is cute; you'd like to call him and arrange a date. But what if he thinks you're boring, unattractive, or too forward? All normal worries and, if you lack confidence, they're paralyzing. You will sit home, nursing a desire to act/call, but not doing anything about it. Confidence propels you to pick up the phone.

Other traits encourage action, as Richard Petty noted. Ambition, for example, which drives us to pursue measurable success, can work in tandem with confidence toward a goal. Courage routinely compels action, is very much inclined to push for action, and early on we almost thought of courage as another confidence cousin. But confidence provides the basic groundwork for action based on a belief in one's ability to do something or succeed, and courage advocates for action with little regard for risk or success, springing from a very different place—a kind of moral center. Courage though, can be a critical partner to confidence, especially in situations where we are operating without the benefit of a confidence reserve, and we need to take those first, terrifying steps in order to start building it.

And sure, other factors can limit us too. Lack of motivation might stop us from applying for that promotion. Procrastination could stop our training for that marathon. But if we assume the desire is there, the only real inhibitor is a lack of belief in our ability to succeed. And, let's be honest, neither the beckoning of a comfortable couch nor a lack of motivation is likely to be what stops us from speaking out at confrontational moments or from cold-calling a potential client to pitch a sale. Confidence is all that matters there.

A couple of questions had been nagging at us, though, since our intense conversations with Cameron Anderson about the merits of overconfidence. What is the optimum amount of confidence? Is that even knowable? With a clear definition of confidence in hand, this seemed easier to address. We had firm agreement from the social scientists and hard scientists on this one—a slight tilt toward overconfidence is optimal. Adam Kepecs, our rat expert, believes it's fundamentally, biologically, useful. "It is adaptive to have appropriate levels of confidence so one makes the right bets in life," he told us. "And, in fact, it is actually adaptive to have a little extra confidence for good measure in the face of uncertainty." In other words—better to believe a bit too much in your capabilities than is called for, because then you lean toward *doing* things instead of just *thinking about* doing them.

You probably have a good gut sense of your confidence level already, especially if you've recognized any of the behavior we've been describing. But there are formal measures. We've put two of the most trusted confidence scales in the notes at the end of the book. One was recently created by Richard Petty and his collaborator Kenneth DeMarree of the University of Buffalo. The other is a thirty-year-old survey still in heavy use. They don't take long, if you want to put some numbers on your current state of assurance.

Confidence, we believe, is our missing link. It's what can propel us out of our overworked minds toward the liberating terrain of action. Confident action can take many forms—it is not always as overt as turning in a job application, or learning to skydive. A decision, a conversation, an opinion formed—those are all driven by confidence.

Confidence, ultimately, is the characteristic that distinguishes those who imagine from those who do. It's the stuff that seems to naturally inhabit the minds of the Susan B. Anthonys and the Malala Yousafzais. But we were also coming to see confidence as something we might *all* create. We recognized an encouraging power in the concept of confidence as action, which, when taken, sows and reaps more of the same. Action, we reasoned, is something we are all free to choose. Might it be that acquisition of confidence is basically our choice? Confirming that appealing notion required answering another question first.

3

WIRED FOR CONFIDENCE

The drive from Washington, DC, to western Maryland brought us to bucolic, red-barn country within an hour. Horses gazed up halfheartedly at our passing car, clearly oblivious to the great experiment going on just down the road. A tribe of three hundred rhesus monkeys, whose original members came from the mountains of Sri Lanka, has made its home in Poolesville. They're here to help humans figure out why we behave the way we do.

We'd come to see the monkeys, and also the man who has been watching them for more than forty years, neuropsychologist Steve Suomi of the National Institutes of Health (NIH). He is a leading explorer of the tangled, centuries-old, nature versus nurture terrain, and he commands a small empire of warehouse labs in this countryside outpost, the centerpiece of which is a five-acre playground for his subjects. The day was gloriously sunny, and many of the monkeys were scampering and swinging on equipment that looked like, well, monkey bars.

"There are truly interesting personality differences in monkeys," Suomi told us. "You see everything from healthy, well-adjusted individuals to monkeys prone to anxiety or depression or even autism. Where do those traits come from?"

Suomi is making huge strides toward answering that question. His wildlife laboratory has become ground zero in the fast-expanding study of the biology of personality.

Specifically, we were on the trail of a confidence gene, wondering whether we could find proof of what was long our gut instinct: some people are just born confident. You know the type, people who appear to glide effortlessly through life, whatever it throws at them. The people for whom no task is too difficult, no situation too agonizing, and no challenge too great. They exude an inherent, enviable, even slightly irritating, air of ease. The professional mother who juggles children, job, and spouse and never questions whether she's doing right by either her family or her career. The young man who sets off backpacking through Costa Rica, just assuming that it will all work out. Those people who have no qualms voicing opinions in public or demanding a raise in private. Their parents, friends, and spouses say they've always been that way, making their seemingly unshakable confidence appear all the more unattainable.

Did their upbringing create that confidence? Or is there a DNA sequence that hatches it? Is confidence baked into our personalities?

Suomi has been asking the same questions and trying to find answers by studying the personalities of his monkeys. Lately, he's been focused on the origins of anxiety, which essentially means, Suomi told us, that he's also looking at confidence. Monkeys with confidence aren't apt to be anxious, and vice versa.

Based on his research and that of others in the field, Suomi has concluded that some monkeys are indeed born with the hard-wiring to be more confident than others. "We now know that there's an underlying biology," he told us. "Certain biological characteristics show up very early in life and, if you don't do anything about the environment, are likely to be fairly stable throughout infancy, childhood, adolescence, and adulthood."

It's an enormous help to Suomi's research that monkeys grow up four times as fast as humans. He's been able to observe several generations already. He and his team track the monkeys' behavior from birth, noting parenting techniques, and marking the frequency with which the offspring socialize with others, dominate the playground, take risks, or hang by themselves.

As we watched the monkeys more closely, with color commentary provided by Suomi and his researchers, it was indeed possible to pick out the different behavior patterns that Suomi described. Some lolled about down by the lake, while others engaged in a game of chase. Several mothers kept watch on their offsprings' every move. We spotted a few of the young monkeys sitting more quietly, close to the adults. One showed almost no interest in even observing the nearby activity. In all, the scene was not unlike what you might see at a grade school playground: Play and interaction dominate, but a few youngsters hang back. Their behavior is typical, Suomi said, for his less confident, more anxious monkeys.

Still, we wondered, as we examined the tableau before us, whether we could really draw conclusions about human confidence based on monkey behavior. Monkeys are our ancient ancestors, and our research had reminded us that we share 90 percent of our genetic makeup. But Suomi explained there's another, even more essential tie between us, one he discovered. The nimble rhesus monkey is the only primate that shares with us a particular gene variation that researchers are coming to see as essential to personality formation. The gene is called SLC6A4, or the serotonin transporter gene, and it directly affects confidence.

We'd heard of serotonin, and you probably have, too. It has a big impact on mood and behavior; more of it can make you feel calm and happy. Prozac and all of its pharmaceutical relatives boost serotonin levels. Serotonin, in short, is good stuff and SLC6A4 is the gene that regulates our serotonin levels by recycling it through our system.

This serotonin transporter gene comes in a few varieties, or in scientific terms, it has a polymorphism, which means it plays favorites; some of us have more efficient versions of it than others. One of the variations is made up of two short strands, which is fairly rare, but people with that genetic hand process serotonin badly, magnifying their risk of depression and anxiety. Another version contains a long and a short strand, which means better, but still inefficient, serotonin use. The third variant contains two long strands, which allows for the best use of the hormone. The people with that variant, scientists believe, are more naturally resilient, which is a key criterion for confidence.

Dozens of studies have examined the SLC6A4 gene in humans. Most have demonstrated clear ties to depression and anxiety disorders and, recently, as scientists have turned toward the study of healthy mental attributes, the gene has been linked with happiness and optimism. Experts like Suomi, who know their way backwards and forwards around the gene, say it's clear that serotonin, especially in its ability to inhibit anxiety, sets the stage for confidence.

Years ago, when he saw the early research, Suomi, who had already conducted decades of meticulously documented behavioral studies of his monkeys, started to suspect the serotonin transporter gene might play a role in what he was seeing. He ran DNA tests on the whole bunch, looking for the serotonin gene. The genetics, when cross-referenced with all of his mounds of data, accurately predicted the behavior he had already recorded: which monkeys had been born depressed, more withdrawn and anxious, and which monkeys were more resilient. He'd hit the jackpot.

We gazed around his office, which was jammed with files and decorated with photos of his brood, many of which were vintage eighties and hanging askew. One, in particular, bears an uncanny resemblance to photos of our children we display around the house—Cocoabean, one of the first monkeys born in Maryland, is taking a glorious leap into the pond of the field station, while her furry pal Eric looks on. The careful nurturing, observation and testing that's been Suomi's life and passion for decades has produced results even he could not have imagined. Happily for us, that work offers a new lens on the genetic origins of confidence, even if in humans confident behavior would seem to be more subtle and varied than it is in monkeys. Being on the retiring side himself, Suomi admitted a certain fascination with our topic. Ruddy-faced, mild-mannered, and sporting a comfortable blue cardigan, he turned his bespectacled gaze away from his research and toward us, and started to describe the variations in confidence he has come to observe.

He has found, for example, that the monkeys with genes rendering them more resilient, or less anxious (the longer strands), tend to be more willing to engage with others, to take risks, and to become leaders of the group. In other words, they show more confidence in their actions. His description of the complex social structure of the rhesus is fascinating and includes patterns of behavior that sound suspiciously like office politics. The leaders focus on alliances and occupy the best real estate; their corner office is a corncob case near the pond. They make their power position clear with a silent, open-mouthed stare at minions and challengers. The smartest up-and-comers are obsequious; the most effective genuflections are grimaces with teeth bared or rumps shoved in the air. The monkeys that have the other versions (the shorter strands) of the serotonin gene don't always exhibit behavior as dramatic or as incapacitating as depression. But, right from birth, Suomi has found them to be less engaged, fearful or clingy and, as they grow older, they are the ones that are less willing to play in risky ways. In other words, they appear to be less confident. Interestingly, in some monkeys, anxiety and lack of confidence manifest themselves as hyperactivity and aggression. It happens more often in the males. That sounded similar to our world as well.

The Human Code

So is confidence encoded in our genes? Yes—at least in part. That's the belief of not only Suomi, but also every one of the dozen or more scientists we interviewed. We all enter the world with a propensity for more or less confidence, and the case goes well beyond the serotonin transporter gene we share with rhesus monkeys. "A lot of personality is biologically driven," says Dr. Jay Lombard, one of the founders of Genomind, a pioneering genetic testing company. "It is clearly both nature and nurture, and understanding what genes do to affect the biology of the brain, to create temperament, is something the NIH has now recognized as a priority."

In terms of scale and duration, one of the most compelling studies that links genes and confidence is a project being conducted by Robert Plomin, a renowned behavioral geneticist at King's College in London. While he can't quite replicate the perfectly sealed and studied Suomi-like habitat, he comes close. And he's doing it with humans.

Twenty years ago, Plomin decided to undertake an ambitious study of 15,000 sets of twins in Britain. He's followed them from birth into adulthood, gaining vast amounts of data on everything from intelligence to a propensity for disease to gender roles. Some of those twins are identical, with identical DNA; others are fraternal and share only similar DNA in the way that ordinary siblings do. Twins have long been the most effective subjects for the study of the nature versus nurture conundrum.

In his recent examination of the academic performance of these twins, Plomin decided to take a closer look at confidence, or the faith the children had in their ability to do well. At age seven, and then again at nine, the twins had been given a standard IQ test and they were also tested academically in three subjects: math, writing, and science. Next, they were asked to rate *how confident they were about their abilities* in each subject. Plomin and his researchers also factored in reports from the teachers. Once all of the data had been cross-referenced, the research team was struck by two findings. The students' self-perceived ability rating, or SPA, was a significant predictor of achievement, even more important than IQ. Put simply, confidence trumps IQ in predicting success. Plomin and his team had found in kids what Cameron Anderson had discovered in adults.

The researchers also found that a lot of confidence comes in our genes. They'd separated the confidence scores of the identical twins from those of the fraternal twins, and found the scores of the identical twins to be more similar. Plomin's findings suggest that the correlation between genes and confidence may be as high as 50 percent, and may be even more closely correlated than the link between genes and IQ.

That a personality trait as seemingly amorphous as confidence might be every bit as inheritable as intelligence struck us as pretty far-fetched, until we discovered we'd ventured upon an entire field of study, the genetics of personality, at a remarkably explosive stage. Countless breakthroughs in the field of behavioral genetics and biology over the past decade have created ever more sophisticated ways to examine the mind in action as well as cheaper, more efficient methods to sequence and scrutinize DNA. Hundreds of these studies—involving genes, brain fluid, behavior, and neuroimaging—make a strong case that large chunks of our personality are formed at conception. Researchers have pinpointed genes that influence everything from shyness to motivation to criminal behavior to a proclivity to be a professional dancer. (It's true. More details on dancing DNA in the notes.)

We should make clear that some of the experts we talked with don't agree with Plomin's conclusion that confidence is *half* genetic. They say that broader personality traits—the big five, as they have become known—are accepted to be about 50 percent genetic. Those are openness, conscientiousness, extraversion, agreeableness, and neuroticism. But they would put attributes such as optimism and confidence, which are considered facets of the big five, in the range of 25 percent inherited. We were still surprised. Whether we get 50 percent of our confidence in our genes, or 25 percent, it's a big chunk, more than we would have thought. (It won't be long, we figured, before newly pregnant women will be able to take a quick fetal DNA test to determine whether they should invest in safety locks and padded walls or cuddly toys and books.)

As tantalizing and voluminous as the science is at this point, it's hardly exact. It turns out that sifting through our 20,000 genes is slow going. Nothing resembling a complete or partial genetic personality code yet exists. Remember, in the twenty or so years since genetic research has taken off, the emphasis has been on pathology—physical and mental illnesses—rather than on the genetic building blocks of health and well-being. That's just starting to change. Now, the equally interesting question is becoming: What do the genes of psychologically strong, healthy people look like?

Not surprisingly, intelligence is the positive attribute that has received the most attention. Researchers around the world have already uncovered at least one intelligence gene by comparing DNA and IQ scores. A young Chinese researcher, Zhao Bowen, is looking for another one, his sequencing machine on overdrive, going through DNA samples from the world's smartest people.

Nobody has undertaken a project yet to extract DNA samples from the world's most confident people, and none of the scientists we spoke to believe that there will be just one so-called confidence gene. As is the case with many complex personality traits, experts told us that confidence is influenced by a large number of genes, dozens or more, which create a messy stew of hormones and neural activity. Confidence involves both emotion and cognition. Indeed it has a metacognitive component, because it involves *our knowledge* about our brain at work. In other words, it's not simply about whether we can *do* a task, but whether we *assess* ourselves to be *capable* of doing that task. Even so, scientists are drilling all around the perimeter of confidence these days, as they examine related personality attributes, such as optimism and anxiety. Their work makes it possible to piece together an early, basic formula.

Fuel for Action

Thinking about our own definition of confidence as fuel for action, we decided that the cleanest approach was simply to ask what it is, exactly, that gets one's brain in the right frame of mind to act.

We discovered that there are a handful of neurotransmitters critical to creating that state of being, working as positive messengers in our brains. Serotonin, the same substance that Suomi is tracking in his monkeys, is one of them.

Healthy levels of serotonin in the prefrontal cortex enable us to make more rational decisions, because serotonin helps us remain calm. Our prefrontal cortex is the command center of our brain—it's the home of executive function, rational thought, and decision making. Think of it as our brain's Yoda. When that part of our brain is awash in serotonin, it encourages confidence in our decision making because we feel much less stress.

That's because serotonin also helps to quiet our amygdala, the primitive part of our brain. It is our primal core, necessary for moments when we need to access strong emotions quickly.

Most of those emotions have a negative association, such as the fight-or-flight response, primal instincts that humans needed for survival on the ancient savanna. In the modern-day era, when survival is a less pressing daily concern (even if it doesn't always seem like it), activity in the amygdala highlights psychological threats, and it contributes to depression and anxiety as well. It's the role of serotonin to calm the amygdala and create what neuroscientists call "healthy communication" between the rational and the fear-based parts of our brain.

Oxytocin is another neurotransmitter that directly affects confidence. That surprised us initially. You may have read news reports about what's been dubbed the "cuddle hormone." Scientists say oxytocin affects our desire to hug, to have sex with our partners, to be generous to friends, to share, to make moral decisions, and to be faithful. It's what women are bathed in when they give birth and breastfeed. Men and women get it through lovemaking and exercise. It's a virtuous circle of a hormone: The more you hug, the more oxytocin is produced, and so you want to hug more. A provocative study, which we also detail in the notes, recently uncovered that it even encourages monogamy.

Shelley Taylor, a psychologist at the University of California, Los Angeles, who studies oxytocin and has found it to be heavily tied to optimism, suggests that it's also a crucial part of confidence. She believes that by encouraging more social interaction, and fewer negative thoughts about others and the world, oxytocin paves the way for people to act and to take risks. When you're optimistic, doing things just seems easier. Oxytocin works in the brain much like serotonin does—helping activity in the prefrontal cortex, the center of higher-order thinking skills and executive function, and keeping the easily alarmed amygdala quiet.

Taylor has even pinpointed the OXTR gene, which controls the delivery of oxytocin. As is the case with the serotonin gene, she has identified two versions of the gene. One can lead to weak social skills, more reaction to stress, low optimism, low self-esteem, and less ability to master things, while the other version correlates with more resilient, relaxed, and outgoing behavior. So, even though we can generate new supplies of oxytocin by having babies and by hugging more, some of us are simply born with more of it, and thus may start out with a higher baseline of confidence-enhancing attitudes and behaviors.

And we mustn't forget dopamine.

Dopamine inspires doing and exploring; it is associated with curiosity and risk taking. An absence of dopamine has been linked to passivity, boredom, and depression. Two relevant genes control dopamine: one is known as COMT, and the other is DRD4. Both of these genes come in different varieties. (Are you starting to sense a theme here?)

One version of the DRD4 gene, DRD47R, is the gene that encourages dramatic risk-taking. It's often called the "adventure gene." Think skydivers, or all of those scandal-prone politicians. Extreme athletes often have it, and those investors who seem to thrive on risk. Their bodies crave the bigger dopamine boosts they get when pushing limits.

The COMT gene is often called the "warrior/worrier" gene. It's complicated, but essential to confidence. We were lured into figuring out how it works by trying to guess whether we ourselves are programmed to worry or fight.

One variant of the COMT gene clears dopamine rapidly from our frontal cortex (warrior), one at middling speed (warrior/worrier mix), and one slowly (worrier). Usually, dopamine is a good thing. Having more there for as long as possible is better for concentration. ADHD drugs are all about dopamine. So it makes sense that the worrier gene variation leaving dopamine in our brains longer, leads to higher IQs. Again—less dopamine means those warriors typically have more difficulty trying to concentrate. Here's the COMT conundrum, though. When stress kicks in, our bodies make dopamine fast. It floods our cortex. And *too much* dopamine does not make for better concentration or risk-taking ability; it overwhelms our brains, causing a sort of stress shutdown. Suddenly, the tables are turned. The genetic variation that removes the dopamine the most slowly, in that moment, the worrier variation, is not a good thing, because it contributes to that shutdown.

So, under stressful situations, the genetic benefits and deficits are reversed. This explains why the extremely focused and responsible worker bees can turn around and choke on tough exams or in other high-risk situations. And why more generally low-key personalities suddenly rise to a specific challenge. In fact, they thrive. They actually need some stress to do their best. Think of the star athletes who impress mainly in the heat of the game. Or, closer to home, of journalists who work well only under intense deadline pressure.

We immediately saw a connection to confidence in the way this variant of the COMT gene might encourage paradoxical behavior. There's actually science behind why self-assurance can appear to be situation-specific. You know, lawyers who are brilliant at preparing briefs, but can't fathom arguing in court. Or the marketing executives who can't get motivated to do the routine tasks their jobs require, but then spring into action just before the monthly presentation, pull all-nighters and come up with winning concepts. To some extent, those people are simply built that way.

We were starting to see how all of these hormones lay the groundwork we need in order to experience confidence. When dopamine, which gets us moving, is commingled with serotonin, which induces calm thought, and oxytocin, which generates warm and positive attitudes toward others, confidence can much more easily take hold.

It was at about this point in our research that we both became insatiably curious about our own genetics and started to speculate, pretty unscientifically, about our DNA. We've tried for years to untangle our own operational paradoxes. Why is it that we rebel against deadlines, for example, but need them to produce our best work?

And yet we're also both fairly cautious about being prepared, and inclined to fall into perfectionism. So are we warriors or worriers? Do we have trouble concentrating except under extreme pressure? And how can we have achieved what we have in our careers, and be writing a book about confidence, and still feel anxiety in an interview? Are we short on serotonin? And what would we do with that information if we had it?

Claire figured she is high in oxytocin, and Katty agreed with that self-assessment. "I crave affection, contact, and intimacy, and tend, to a fault, to view the world around me with the rosiest of lenses. It can cloud my judgment sometimes," Claire admitted. "But I'm guessing I'm low in serotonin, because I can be quite anxious. I've struggled with it over the years, and my parents both suffered from depression. I think my overall makeup would not add up to a highly confident genetic profile. I've probably created whatever confidence I have."

We were both certain that Katty is a warrior. From Claire's point of view especially, Katty thrives on risk and challenge. "I figure I'm also pretty high in serotonin—I'm not overly anxious," says Katty, "but I don't think I have the warm and fuzzy oxytocin gene. I'm fairly matter-of-fact."

We went back and forth in this vein for a few weeks, debating a move that neither of us had anticipated when we started researching this subject—whether we should seek out a genetic mapping for ourselves. We hadn't expected to uncover much research to suggest that confidence could be inherited, or biologically based, but now we were hooked on the idea that we might be naturally, genetically, inclined to jump off a cliff or to stand up in a really intimidating meeting. Would it help us to know? Or would it inhibit us further? Ultimately, curiosity won out and we decided to do the tests, especially when we found out that it's as simple as sending off saliva samples to one of two genetic start-up companies, one called 23andMe, the other, Genomind.

23andMe (named for the number of chromosome pairs in each human cell) is the Google-backed personal genetics company that suddenly started generating headlines over a showdown with the U.S. Food and Drug Administration just as we were finishing the book. In late 2013 the agency warned 23andMe that it could not market its genetic tests without regulatory approval, and the company has stopped selling them for now, as it attempts to negotiate a blessing from the FDA. The ongoing debate promises to help define the new era of personal genetics and personalized medicine, and most especially who controls access to DNA sequencing and information.

23andMe had been offering something unique—broad genetic testing, directly accessible to the public, without going through doctors. It had become fairly affordable—they were selling their personal genetics test for just ninety-nine dollars. The company has declared its long-term goal is to become a leading genetic database for scientific research. Its testing wasn't comprehensive—a million out of our three billion genes. But 23andMe scientists were focusing on genes where they believe research has identified health concerns—genes that affect Alzheimer's, Parkinson's, and breast cancer, for example, and quite a few others. They were also offering a lot of specific information about ancestry. The results included substantial detail about what your genetic profile says about a broad array of other potential health risks. And that, essentially, is what troubles the FDA, which contends that there is conclusive science for only a very few genes right now, and that consumers might, therefore, put too much or too little weight on their results and make questionable health decisions, without the help of their physicians.

Genomind's testing is more narrowly focused, but it's revolutionary in the field of brain wellness and neuropsychiatry. The founders want to put cutting-edge genetic science in the hands of doctors and psychiatrists for concrete treatment. (The company has avoided issues with the FDA by offering testing only through physicians and hospitals.) Genomind scientists have created a very specific panel of genetic tests, backed up by analysis of results, which healthcare providers around the country are already using. Instead of a doctor listening to a patient describe symptoms, and then using a trial-and-error approach to medication, the genetic results often suggest which medications might be most effective. You may be a candidate for antianxiety medication, for example, but if you have a certain genetic profile, some drugs won't work well for you. Your doctor can choose, based on the science, to start with something else.

Both 23andMe and Genomind offered some, but not all, of the testing we were looking for, and 23andMe was still in full operation then, so we doubled down. We understood, by that point in our research, that genetics aren't determinative. But still it felt like an important decision.

The test taking was almost too easy (we each spat in a test tube), but anticipating the results was agonizing—a bit like waiting for your SAT scores to come through. As reporters, we always figure that any information is good information. Knowing our raw material had to be an asset. That's the very point of this chapter. But, what if? What if we didn't like what we learned? What if the results reinforced all our negative self-stereotypes? No ruminating, we told ourselves. We'd just have to wait.

What Happens When Life Happens

If a large chunk of confidence can be explained away by genetics, we wondered what that meant for our theory that confidence might also be a choice. It turns out both are possible. We discovered that for a lot of the scientific world, the long-standing nature *versus* nurture formulation is old news.

The pioneering thinking and research has moved on to examine the frisson that happens when nature and nurture interact. It's the effect of nurture on nature that really matters and makes us who we are. In many cases, nurture is so powerful that it can alter nature's original programming, turning genes on and off, as it were. Some scientists are uncovering "sensitivity genes": genetic variations that may mean those who carry them are more susceptible to environmental influence than others. Still other researchers have found that the power of habitual thinking creates physical changes and new neural pathways in our brains, which can reinforce and even override genetics and change brain chemistry, as well. So, life choices do matter, as much as, if not more than, what we're born with.

Think about it like this. You have a blueprint for your new house, and the foundation has been poured. Some structures will be easier to build on top of that foundation than others. If you're lucky, you already have the underlying support for a third story. But, even if your foundation isn't as strong, and you have to pour a bit more cement, you can still add that additional floor later on. It might take extra work. And you might have to use different materials. Much will also depend on outside factors. How much will storms batter the house? Are you in an earthquake region? Or are you blessed with mild weather? Those weather and geological conditions will force the foundation to move and react in different ways than it otherwise would. However, your time and effort matter as well.

Confirmation that this is the new-new direction in science came when we saw a very real structure, a half-billion-dollars' worth, going up double time on the Upper West Side of Manhattan. Our journey brought us to the front of what will be the home of Columbia University's Mind Brain Behavior Institute. Its goal is to create a holistic approach to the study of the brain, its functioning, and its impact on everything from human behavior to health to emotion. Scientists and psychologists but also historians, artists, and philosophers—top scholars from every field—are converging on this Columbia outpost in West Harlem.

The thinking is that this merged approach will bring answers faster and begin closing the gap between science and behavior. After countless uncomfortable hours of writing in our small home offices and at multiple Starbucks, of conducting frazzled interviews on trains, and reading research in cabs, we enviously examined plans for the institute by award-winning architect Renzo Piano, who specializes in buildings meant to inspire lightness and creativity. This one will give the appearance of floating above the ground and will have a vertical core of open space at the center of each floor designed to encourage interaction and brainstorming. Labs will be standing at the ready when ideas blossom.

The institute's codirector, Tom Jessell, a professor of biochemistry and molecular biophysics, used to spend his days studying the mysteries of the microscopic. Now he thinks irrepressibly big. He sees around the corners of confidence in ways we haven't even imagined.

For Jessell, understanding confidence from the cellular perspective, all the way out to its global implications, is the only proper scientific path of study. He told us to think about parts of the world where people lack a sense of individual power. He leapt up from his chair, his excitement uncontainable. "What you see may be, in part, the consequence of not having confidence. You live in a world where everything you do is bad and nothing you do makes a difference. It's what's called 'learned helplessness.' There are social scientists in Africa studying that." Of course poverty, climate, and bad governments play a crucial role, he said. "But if, through multidisciplinary study, you can come to understand what you can do to change that state of confidence in an individual, how you can move from a state of apathy to one of optimism, the global implications would be amazing."

Scientists at Columbia and at other leading universities are at the forefront of bringing the macro and the micro together through a revolutionary field called *epigenetics*, which examines how life experiences can become imprinted on our DNA and can change the *epigene*, or the outside of our genes, and cause the genes to behave in different ways.

Certain traits are pretty fixed and hard to influence—qualities such as height or eye color, for example. But character traits, like confidence, are much more complex and malleable. Identical twins offer the best example of the power of epigenetics. True, their DNA is the same, yet they often show differences in health and personality. Why? It's due to the *expression* of those genes, the way some are switched on while others are switched off. Those on and off switches are heavily influenced by the external environment.

Even more profoundly, those external changes may be passed on to our children immediately. Genetic change may be possible in a single lifetime, instead of over multiple generations, as Darwin proposed.

"The whole idea of whether our lifespan experience can become heritable is an incredibly hot topic in epigenetics right now," said Frances Champagne, a psychologist at Columbia who specializes in the field. She and her researchers are exploring the ways in which prenatal events can induce long-term effects. Her findings suggest there are various ways that stressful episodes, for example, can have an epigenetic effect not only on pregnant women but also on their developing fetuses. She's also looking at the impact of stress on men and their future offspring. Other studies have found that women who witnessed the 9/11 attacks while they were pregnant passed on significant levels of stress hormones to their babies, through their DNA. A different study found that pregnant mice that got fewer vitamins had babies more likely to be obese. It's entirely too early, Champagne told us, to predict whether a woman who builds her confidence, for example, might at the same time be creating something heritable for her children, but it's not out of the question.

Strength Through Sensitivity

Steve Suomi has been in a unique position to study some of the new trends in the nature/nurture puzzle. Suomi can play Svengali with his monkey colony in a way in which no psychologist researching humans would be able to. He has taken monkeys that are anxious or confident, based on their genetics, and then carefully manipulated their environments to see what happens. The results are startling. "Some traits are inherited," he said, with a slow smile, "but that doesn't mean they can't be altered."

Suomi has found that, just as with humans, rhesus monkey mothers are extremely important in shaping the attitudes and behaviors of their offspring; the first six months of bonding and nurturing are critical. Just how critical? "We did studies looking at monkeys whose genetic background suggested that they would be naturally anxious and fearful, and we cross-fostered them with mothers who were supportive and there for their kids," he told us. "And those kids did beautifully when they grew up. They grew highly social. They got used to looking to others for help, and they ended up at the top of their dominant cycle."

So here's how it breaks down: Monkeys that were born with the more resilient genes essentially did fine with any type of mother. The monkeys that have the social anxiety gene, raised by anxious or neglectful mothers, grew into anxious adult monkeys. A decent mother produces a somewhat anxious adult, but a great mother can turn a baby genetically programmed to be at risk for anxiety into a healthy adult. With her nurturing, her child can overcome the genetic blueprint.

Suomi then made a radical, counterintuitive discovery with far greater implications. Those so-called "genetically challenged" monkeys, when raised by those great mothers, don't just turn out fine, they actually excel. They thrive. They become stronger, healthier, and more confident than their peers. They become superstars, if they have superstar moms.

Suomi had uncovered what a few other researchers are starting to understand. Some genes make monkeys, and humans, not more *vulnerable* to the environment, but more *sensitive* to the environment. There's a big difference. Suomi has come to see the monkeys with the anxious gene as sponges, absorbing the worst, but also the best, of what they experience.

In scientific circles, the proposition of sensitivity genes is quickly gaining ground and was recently dubbed the orchid theory. Most children, according to developmental psychologist Bruce Ellis and developmental pediatrician W. Thomas Boyce, are genetically like dandelions: hardy and able to thrive in many environments. They go on to suggest that despite what we have thought for years, the nondandelion children may in fact not be the weak ones. Based on burgeoning evidence, the researchers posit that these children should be viewed as orchids: trickier to raise, but if nurtured in the right environment, able to excel beyond even their sturdier dandelion counterparts.

Other studies confirm that for people with sensitivity genes, the environment leaves a much larger imprint. Thousands of German toddlers, whose routine screaming, whining, and unfocused behavior was so severe that they were considered at risk, were followed recently for almost two years. Researchers videotaped parental interventions, and then guided the parents as to how to better interact with and read with their children. At the end of the study, the researchers noted significant behavioral improvement in all of the children. The most encouraging surprise? The biggest improvements came for the children with a version of the dopamine gene that has been linked to ADHD, but may actually be another sensitivity gene. For them, the positive parental interventions produced twofold improvements over the children with the normal gene who received the same kind of interventions.

Scientists at the University of Essex tested this theory on adults, using a computer game with a variety of images, and discovered that people who have the short forms of the serotonin gene are more easily influenced by both negative and positive information. They are more sensitive, and, some scientists believe, more able to adapt.

Think of it this way, in terms of the real world. The people born with this gene might be among the least confident you know, or among the most confident, depending on the challenges they've faced and the support they've had growing up.

"And of course," noted Suomi, with a shrug, "it's a roll of the dice as to which environment you find yourself in." He believes that there are other critical periods, beyond childhood, when those with sensitivity genes might be especially influential. He's started to study the relationship between the serotonin gene and puberty, childbirth, and menopause.

The orchid gene hypothesis, by the way, fueled our nerves, waiting for our genetic results. Were we dandelions or orchids? Katty guessed she was of sturdy stock, able to adapt and cope in many environments. "But you're so flexible," said Claire, who was now thoroughly confused about what her own results would show. Did her ubernurturing mother play a big role? Or was her resilience in her DNA? She started to obsess about her kids. "Who needs extra attention? Should I test them? And if I have an orchid, would that child suck up all my mental energy, leaving me little left for the other?"

The Promise of Plastic

The most astonishing science, with profound promise for cracking the confidence code, proves that all of us can rewire our brains, even as adults. Orchid or dandelion, good mother or bad—when we change our thinking and develop new mental habits, that effort creates physical changes in our brains.

The question of resilience, of why some people are better at withstanding setbacks than others, why they remain confident in the face of disasters, has preoccupied Rebecca Elliott for years. A leading researcher in cognitive brain imaging at the University of Manchester, she looks for clues in those images about resilience. Resilience may be partly genetic, perhaps a result of that serotonin gene. But Elliott believes research will soon bear out that resilience, a quality related to confidence, can also be created, and she pointed us to the expanding research on brain plasticity. Fairly simple brain training, she explained to us, or methods of thinking, can carve new pathways in our adult brains, pathways that encourage resilience, or confident thinking, and that then become part of our hard-wiring.

It may have been when we started to understand that there is just so much new data on brain plasticity that we could never get through it all, or it might have been when we finished yet another riveting decoding session with Laura-Ann Petitto, a groundbreaking cognitive neuroscientist at Gallaudet University—it's hard to pinpoint the moment, exactly. But fully understanding the promise of brain plasticity was a watershed for us. It's fair to say it changed our outlook on the project. There had been times, as we examined the causes of the confidence gap, when we were weighed down by the thought that it might take another few generations to overcome. Indeed, for that reason, we weren't planning on this as an advice or how-to book. We just didn't think we'd find a lot that was truly relevant, beyond the superficial dictate to sit up straight. We'd also been disturbed to learn that a good bit of our self-assurance is genetic and were starting to wonder how much of a choice some of us have in the matter. We had no idea something as straightforward as targeted mental exercises could create lasting behavioral changes.

Plasticity is the cornerstone of the idea that confidence is a choice we can all make. If we can permanently alter our brain makeup, then even those people born with less genetic confidence can develop permanent, solid confidence with the right training. Norman Vincent Peale was far ahead of his time. There's power *and* science in positive thinking.

(The more we read about plasticity, the more it calmed our anxiety about having commissioned those genetic tests. At least, we figured, we could overcome whatever unsavory results turned up.)

Like most parents, we'd been familiar with the concept of brain plasticity in relation to our children's brains—the idea that we needed to shove in a bunch of good stuff before they turn ten and their brains supposedly turn rigid and brittle. Actually, the window is open much longer than that. Our brains remain plastic throughout adulthood.

Elliott told us that cognitive behavioral therapy, a technique developed to help individuals create new thought patterns, is the most effective approach to making specific behavioral changes, but some of the most dramatic examples of a change in the brain's function and structure have involved basic meditation. A number of studies, conducted with MRIs (magnetic resonance imaging) before and after a period of meditation, showed less activity in the amygdala, the fear center, after an average of eight weeks of meditation. A recent experiment, done with highly stressed business people, found not just less fear activity after meditation, but also that the amygdala itself actually shrank and remained smaller. Conversely, those same postmeditation MRIs show more activity in the prefrontal cortex, the center of calm reason.

We had both tried meditation and knew that it made us feel calm. Now, understanding that it could physically change our brains made us resolved to make the practice a regular habit.

Beyond meditation, clinicians are having surprising success treating post-traumatic stress disorder victims with cognitive therapy. PTSD typically shifts action in the brain to the amygdala. And cognitive behavioral therapy moves it back to the frontal cortex. Researchers at Northwestern University documented remarkable changes in the brain's physical makeup after a short session of behavioral therapy aimed at patients battling a fear of spiders. They studied twelve adults with arachnophobia. Before therapy, the brain scans showed the regions involved in fear, especially the amygdala, had much stronger responses to spider photos. They then had a single two-hour session of behavioral therapy. In this case, the therapy involved approaching and then touching a live tarantula. (Talk about facing your fears. It's like asking someone with a fear of public speaking to practice with a podium and an audience.) When they repeated the brain scans at the end of the treatment, the amygdala's action was back to normal. But the prefrontal cortex, the part of the brain responsible for reappraisal and for looking at things more rationally, was *more* active. And here's what's most impressive—when the subjects' brains were scanned again six months later, the amygdala was still quiet. *After just a two-hour session.* Six months later, they were still able to calmly touch that tarantula.

What else happened in those therapy sessions? The participants were taught that many of their fears about spiders were unfounded. Some thought spiders might be able to jump on them. Others thought tarantulas were planning something evil. (You can see how this might work in the case of talking through fears about a difficult boss or about coworkers.) They were taught that the tarantula itself was most interested in hiding from humans. Essentially, they learned to keep their catastrophic thoughts in perspective.

Cognitive therapy is a conscious focus on creating changes in our brains. Another element that affects brain plasticity, of course, is all the stuff that we store and use unconsciously. Memory, that repository of life experience, is a huge player in confidence. Think about it. The past is always prologue in our brains. Memory is one of the elements that make our confidence mechanism far more complicated than it is for rats. The way we interact with our environment is based on a preconception of what the world will do to us, which is based on memories of past experiences. We play and replay that tape in our heads.

More important, we may be playing that tape without even being aware of it. Daphna Shohamy, a Columbia neuropsychologist, used an fMRI (functional MRI) machine to scan the brains of students while they played a series of video games. One simple game involved subjects choosing between two images, after being told that certain images came with rewards. Later, while playing a different game, subjects were asked to choose randomly between two non-rewarded images. They had no understanding that this was happening or a sense of the pattern that was involved, but the picture they tended to pick was the one that had been shown *next* to the rewarded choice in the earlier round. The association seemed to have rubbed off and to have been stored and accessed in their brains. The MRIs validated that, because they showed the hippocampus, or memory home, lighting up in that second round, suggesting memory had been accessed in making that choice.

Yet when the students were asked later why they made their choices, they had no *conscious* memory of choosing based on that reason. It's the first time that anybody has demonstrated that the hippocampus, a hefty middle part of our brain, does more than just consolidate memories. It muscles its way into our cortex as we make choices, pushing and prodding, but leaving no prints.

We intuitively understood how this might affect confident action. Our memories, conscious or not, are informing what we decide to do next. A memory of a negative comment from a colleague in a meeting four years ago may still be contributing to our tendency to keep quiet. Conversely, a few successful speeches in college, even though we may no longer remember those experiences, may be giving us the confidence to speak at the company's annual meeting.

We both started wondering about these hidden but influential happenings in our lives. Katty thought back to a time when, as a young reporter, she completely fumbled a live report from Japan. She couldn't help wondering whether, all these years later, that memory resurfaces whenever she feels a bit of nerves on live television. Claire remembered a rejection for a part in a high school play and wondered about the impact that had on her conclusion that she didn't really like acting much.

Obviously, no one can escape rejection or avoid the pain of a botched performance. We can't fully control the experiences that will eventually become searing and unconscious memories. But just knowing how heavily our unconscious memory can weigh on future action means we have to build up plenty of positive alternatives, because they will matter.

Laura-Ann Petitto says brain plasticity is the biggest neuroscience breakthrough of the decade. "Suppose," she told us, "some of your lack of confidence comes from a more Freudian, less genetic place. It may well come from patterns created in childhood, based on how your parents treated you, or on how others perceived you. Your neural tracks will lay down memories in response to that. Think of it as a cement highway that can create knee-jerk first responses in the future. But, if you can layer that over with new memory networks, you can reroute the highway. You can build bridges over the highway. You might not get rid of the highway, because it was put down so early, but you can work around it, and literally lay down new roads." It's a remarkably effective way to break a key confidence killer: negative habitual thought. What's most startling, Petitto says, is the extent to which it's now possible for neuroscientists to *see* these changes, to watch the brain rerouting, and to witness the new neural routes forming.

UCLA psychologist Shelley Taylor already sees broad prescriptive benefits from understanding our biology. "The potential for the environment to interact with the gene is certainly higher than people have estimated. You might have the version of the oxytocin gene that would predict shyness, but supported in the right fashion, by parents and friends and teachers, you might not even consider yourself shy. However, if your parents are also shy and retiring, and you aren't encouraged by teachers and friends to engage, you will likely follow the path the genes have laid."

Petitto agrees that environment modifies predispositions, but not totally. From birth, for example, some babies are labeled "high arousal," and some "high attentive." High attentives are kids who can usually stimulate themselves; they tend not to be bored. They are often more confident, because they don't need a high degree of external validation to tell them that they are competent.

High arousal babies, however, are often those who are inconsolable and demand attention. They can become the teenagers who end up in trouble. "As kids, they are the ones who constantly go for danger, who find high-risk danger delicious. They can be raised by nuns but still grow up and look for danger," Petitto says with a laugh.

Yet again, we found ourselves focusing on our daughters, who sometimes behave just like boys (we know, a cliché), operating with little caution and no fear of consequences. Wiping dirty putty on the living room wall, for instance, to see whether it leaves marks (yes); riding a sled down the steps to see what crashing will feel like (painful); or disregarding our warnings about the repercussions of launching an indoor water fight (no television for a week—aggravating for all involved).

With these extraordinary leaps in science, we began to channel Orwell and imagine a time when we might be able to learn our genetics, and then know how to structure our environment to become our best selves. We wondered whether we should find out about our kids' genetic mix. Would we handle them differently? It was a step too far; we each knew we were not ready for that kind of information. We were nervous enough waiting for our own.

The state and pace of neuroscientific research and genetic discoveries made for an exhilarating part of our exploration. We'd come to understand that while we get a confidence framework at birth, we can alter it significantly. We do have a choice in the matter. We realized, though, that there was one thing we had *not* uncovered: a genetic smoking gun. We'd seen no conspicuous evidence in our inquiry thus far that men have proprietary access to any sort of master confidence gene; nothing to clearly and tidily explain an imbalance of confidence. Nurture, and environment, demanded some close examination.

4

"DUMB UGLY BITCHES" AND OTHER REASONS WOMEN HAVE LESS CONFIDENCE

The young men at the U.S. Naval Academy in Annapolis, Maryland, have a name for female students. They call them DUBs—dumb ugly bitches. Yes. Disgusting. We didn't quite believe it until we verified it with a number of recent graduates. The men blithely insist it's a term of endearment, and indeed, it has now become so pervasive that even some of the women use it, too. Imagine living and trying to rise to the top in an environment where that's what you're called.

Navigating the Naval Academy can be tricky for a woman, and there are plenty of things Michaela Bilotta chose to let roll off her back while she was there. DUBs was not one of them. She hated the phrase, and made a point of telling people to pick another term when they were in her presence. But she also knew that she had to survive four years at Annapolis, so when she did let people know that she disapproved, she tried to be polite.

What we heard in that coarse language being used at one of the most respected institutions in the country is more than an ugly slur. It contains echoes of the centuries of imbalance that accounts for some of the confidence gap between men and women today. Genetics help explain why some people are naturally more confident than others, but it doesn't sufficiently account for the gender difference. We wanted to know what women are doing to themselves—or what is being done to them by others—that might shed some light on the confidence gap.

The atmosphere at the U.S. Naval Academy definitely falls into the "being done by others" category. Of course, it is an extreme example, but if you ever need proof that women, like Ginger Rogers, are still dancing backward and in high heels, remember that term DUBs. When women have to deal with abuse like that, it's little wonder that so many of us struggle with confidence.

It's been half a century since women first forced open the boardroom doors, and the workplace terrain still looks very different for us than it does for men. The statistics are well-known, and they aren't pretty. Women earn on average 77 cents for every dollar earned by a man. Four percent of CEOs in the *Fortune* 500 are women. Twenty of the 100 United States senators are women, and even that is celebrated as a record high.

We now know this discrepancy isn't caused by lack of competence. Over the past fifty years, women in the United States have reversed the education gap and turned it in their favor, now earning more undergraduate degrees, more graduate degrees, and even more PhDs than men. A half dozen global studies, from Pepperdine to the IMF, now show that companies that do employ women in large numbers outperform their competitors by every measure of profitability.

When women are given a fair shot at success, they do well. Take the intriguing case of classical musicians. Back in 1970, women made up only 5 percent of the musicians in America's top symphony orchestras. By the mid-1990s, we were at 25 percent. The gains came after orchestras introduced a remarkably simple change in how they chose their new hires. During their auditions, they put up a screen to hide the candidates' identity. The judges heard the music, but they couldn't see whether the performer was a man or a woman. Based exclusively on the sweet sound of performance, women began getting hired in greater numbers.

From Annapolis to the New York Philharmonic, some of the reasons women lack confidence can be found in our environment. Sometimes the inequities are outrageous and obvious. Often, though, the cards are stacked against us innocently, with the very best of intentions.

"If Life Were One Long Grade School, Women Would Rule the World"

Take a trip down memory lane to your elementary school classroom. There you'll find the insidious seeds of society's gender imbalance, because it's there that we were first rewarded for being good, instead of energetic, rambunctious, or even pushy.

It is in school that girls are expected to keep their heads down, study quietly, and do as they're told. We didn't charge around the halls like wild animals, and we didn't get into fights during recess, and today's girls still provide a bit of reliable calm behavior for overstressed, overworked, and underpaid teachers. From our youngest years, we learn that cooperating like this seems to pay off.

Peggy McIntosh, associate director of the Wellesley Center for Women, thinks that encouraging our girls to be compliant can do real long-term damage, but she also thinks that it's hard to avoid. It's actually easier for young girls than young boys to behave well, because our brains pick up on emotional cues from an earlier age. So we do it because we can, and then because we're rewarded for it. We do it for our teachers and our parents, too. Soon we learn that we are most valuable, and most in favor, when we do things the right way: neatly and quietly. We begin to crave the approval we get for being good. And there's certainly no harm intended: Who doesn't want a kid who doesn't cause a lot of trouble?

The result is that making mistakes, and taking risks, behavior critical for confidence building, is also behavior girls try to avoid, to their detriment. Research shows that when a boy fails, he takes it in stride, believing it's due to a lack of effort. When a girl makes a similar mistake she sees herself as sloppy, and comes to believe that it reflects a lack of skill.

This is not a message that Claire's daughter seems to have taken to heart, thank goodness. Della is nothing like her perfectionist, teacher-pleasing mother. She's a tomboy, absolutely fearless. She despises dresses and looking neat and doing anything with her hair, some of which she herself chopped off recently.

"It's challenging sometimes, to say the least, having a daughter who doesn't conform to society's expectations—people don't encourage girls to be grimy and loud and obstreperous. But I had a breakthrough moment with her the other day, when I realized she'll do well in life, as long as I don't mess with her natural path." Claire recalls, "I'd been encouraging her to raise her hand in class—to participate. I asked her when she came home that day whether she'd raised her hand. 'Yes, Mom, I did,' she said. 'In fact I raise my hand all the time now, even when I don't have anything to say.'

"Initially I wanted to offer some motherly advice about 'being prepared,' and then I thought, thanks to all of the research we'd done, 'How terrific! What a metaphor for confidence, for all of us. How *masculine* to think about raising your hand, even if you have nothing to say.'

I'd since used this gem of an anecdote often, to great laughs and appreciation, and so was surprised when Dr. Richard Petty, who'd so kindly took the time to read our entire manuscript and offer valuable analysis, said that my use of her story as general guidance was the only piece of advice in the book he thought was misguided. Taking seriously foolish risks, such as volunteering yourself to speak publicly when you don't actually have anything to say, could easily have devastating confidence consequences, he pointed out. I thought about it. Indeed, he's right. Thank you Richard! I realized I had been somewhat glib in so thoroughly embracing Della's gumption regarding risk-taking. Truly having nothing to say, and yet attempting to speak up, can clearly backfire. So, after further thinking, let me clarify why I still, instinctively, can't let go of the power of the image of female hands spontaneously popping up, in all walks of life. I also believe that Della, and most women, need to realize that while we may *feel* as though we're raising our hands with 'nothing to say,' thanks to our still faulty confidence meters, I'm betting that we usually, in fact, have plenty to contribute, and will surprise ourselves. I suppose I am quite confident, actually, that once we start to nudge our hands skyward, we will find our wisdom to be easily unleashed."

Della's style is an exception. Most of us learned the good girl lesson all too well. But it doesn't prepare us very well for the real world. Carol Dweck, author of the best seller *Mindset* and a Stanford psychology professor, puts it this way: "If life were one long grade school, women would be the undisputed rulers of the world."

Owning the Classroom, Skipping the Playground

The meritocratic academic classroom, where we excel, doesn't teach us to play very confidently in the assertive, competitive world of the workplace. With all their focus on getting high academic scores, too many girls are ignoring the really valuable lessons that wait outside of school. "Girls still don't play enough competitive sports, where we train them to know what it's like to compete and win," says Susannah Wellford Shakow, cofounder of Running Start, the group that prepares women to run for political office.

We all know that playing sports is good for kids, but we were surprised to learn just how widespread the benefits are. Studies evaluating the impact of the 1972 Title IX legislation, which made it illegal for U.S. public schools to spend more on boys' athletics than on girls', have found that girls who play team sports are more likely to graduate from college, find a job, and be employed in male-dominated industries. There's even a direct link between playing sports in high school and earning a higher salary in later life. Learning to own victory and defeat in sports is a useful lesson for owning triumphs and setbacks at work.

The number of girls playing sports has increased dramatically since Title IX passed. In college, women's participation in athletics rose sixfold from 1972 to 2011. In high school, girls' participation in sports jumped a staggering 1,000 percent over the same period. But the numbers are still uneven. Fewer girls than boys participate in athletics, and many who do quit early. The Centers for Disease Control and Prevention is alarmed that girls are still six times as likely as boys to drop off their sports team.

Academics confirm what we know from our own experience as teenagers: girls suffer a larger drop in self-esteem during adolescence than boys, and it takes them longer to get over those demoralizing years. The drop in confidence makes them more likely to quit team sports because their self-confidence isn't robust enough to handle losing. What a vicious circle: They lose confidence so they quit competing, thereby depriving themselves of one of the best ways to regain it.

Men, meanwhile, seem to more naturally embrace competition, whether it's for the boss's attention, their peers' adoration, or the sweepstakes for the corner office. Out on the football field, boys learn to relish wins and flick off losses. In the classroom, boys tend to raise their hands before they've even heard the question, let alone formulated a reply. Essentially, they turn everything into a competition. All of this behavior may irritate the teacher, but it's hard not to envy that degree of confidence.

With all their roughhousing and teasing, boys also toughen each other in ways that are actually useful for building resilience. Where many women seek out praise and run from criticism, men usually seem unfazed, able to discount other people's views much earlier in life. From kindergarten on, boys tease each other, call each other slobs, and point out each other's limitations. Psychologists believe that playground mentality encourages them later, as men, to let other people's tough remarks slide off their backs. It's a handy skill to have when they head out into the cold world.

Girls leave school crammed full of interesting historical facts and elegant Spanish subjunctives, so proud of their ability to study hard and get the best grades. But somewhere between the classroom and the cubicle, the rules change and they don't understand it. They slam into a world of work that doesn't reward them for perfect spelling and exquisite manners. The requirements for success are different, and their confidence takes a beating.

Professional success demands political savvy, a certain amount of scheming and jockeying, a flair for self-promotion and not letting a *no* stop you. Women often aren't very comfortable with that. Perhaps, deep down, we don't really approve of these tactics. Whatever the reason, we haven't been very good at mastering these skills, and that holds us back.

Valerie Jarrett regularly spots this operational tension in the women she works with. One of the top women in the White House, she's a senior adviser to President Obama, and also an unofficial adviser to dozens of female White House employees. She's an especially effective messenger because she freely admits she's worked hard to banish her self-doubt. We dropped in late one afternoon, and, along with some female colleagues, gathered around a conference table in her West Wing office. In a cream silk blouse, splashed with an edgy design of purple and yellow (she's known as a sharp dresser), Jarrett managed to convey crisp authority and matriarchal warmth at the same time. We noticed, over the course of the hour-long conversation, that Jarrett listens and solicits other opinions as much as she speaks, even in a session in which *she's* being interviewed. One thing she's learned over the years, she told us, especially watching her friend Tina Tchen, the First Lady's chief of staff, is that you don't always have to dominate a conversation to have an impact.

But there are times when speaking up is required, and women have got to master that distinction. "We're taught to be more self-deprecating," she told us. "I think it all begins on the playground, and then society reinforces it. We believe that we should wait until we are absolutely sure that we are ready for something before we ask for it." It took her a decade in the workplace to learn to ask for something boldly, without waiting. She was in her early thirties, working in the Chicago mayor's office and doing a stellar job handling huge real-estate transactions. A client told her she was doing the work her supervisor should be doing. "She told me, 'You need to be the boss. You need a promotion.'" Jarrett didn't believe it. "I thought she was crazy, but she kept nudging me," she said wryly, "for months and months and months." Eventually, she listened, and decided to take a chance and just ask her boss. She remembers the meeting as though it happened yesterday. "I was so nervous, but I told him all of the reasons why I deserved it, and he, very quickly, just said, 'OK.'" It was as though the scales dropped from her eyes. Emboldened, she asked for a front office, too. He waffled, and a few days later she simply moved in to an empty one. It was a breakthrough confidence moment for her.

Years later she asked her former boss, now a good friend, why he never just offered her a promotion. He told her he'd been busy, and hadn't thought about it. "We all assume," Jarrett told us, "that there's a reason why. We think, 'I'm not deserving, if I were, he'd recognize my talent. It's not up to me to point it out.'" It's a way of thinking she sees routinely, even in the White House, and it's one she tries to vanquish because she knows the career damage it can do.

Consider the following tale of two employees working in New York. A female friend of ours was supervising the two twenty-something junior staffers, one female (whom we will call Rebecca) and one male (whom we will call Robert). Even though Robert had been in his job for only a few months, he was already stopping by our friend's office to offer off-the-cuff pitches for new ad campaigns, to comment on business strategy, and to share his unsolicited opinions about recent articles he'd read in the *Economist*. Our friend often found herself shooting down his ideas, correcting his misperceptions, and sending him off for further research. "No problem," seemed to be his attitude. Sometimes he'd retort with a counterargument; other times, he'd grin and shrug his shoulders as he headed back to his desk.

A few days later, he'd be back in to pitch more ideas and to update her on what he was doing even if it was only to say, "I'm still working on that." Our friend was struck by how easily and energetically Robert engaged her, and how markedly different his behavior was from Rebecca's, with whom she'd already been working for several years. Rebecca still made appointments in advance to speak with her and always prepared a list of issues and questions for their discussions. When asked to provide feedback, she did, but she was mostly quiet in meetings with outside clients, focusing instead on taking careful notes. She never simply blurted out her ideas; she always wrote them up with comprehensive analyses of the pros and cons. Rebecca was prepared and hardworking, and yet . . . even though our friend was frequently annoyed at Robert's assertiveness, she couldn't help but be impressed by him. She admired his willingness to be wrong and his ability to absorb negative feedback without letting it discourage him. Rebecca, however, took negative feedback hard, sometimes responding with tears, and a trip to her own office to collect herself before the conversation could continue.

Our friend had come to rely on and value Rebecca, but when she speculated as to which of the two had what it takes to go far, she knew it was Robert's star that would rise. It was only a matter of time before one of his many ideas would strike the right note, and he'd be off and running—probably, our friend was beginning to fear, while Rebecca was left behind, enjoying the respect of her colleagues but not a higher salary, more responsibilities, or a more important title.

Faced with these corporate realities, sometimes we women give up altogether, deciding we don't fit in this world and can't be bothered to put up with it when the toll on our psyches and our families is so high. Too often, even when we stay, doing so drains our energy. Every morning we have to drag on our office armor, trying to win a game we don't really understand or like.

Same Game, Different Standards

Here's an unsavory question: If Rebecca did behave just like Robert, exhibiting his kind of self-assuredness, what would her boss think then? All the evidence suggests that Rebecca wouldn't fare so well, whether her boss was male *or* female.

For women, this is the big conundrum of confidence. A host of troubling studies now show that we pay a heavy social and even professional penalty when we act as aggressively as men do. If we walk into our boss's office with unsolicited opinions, speak up first at meetings, and give business advice above our pay grade, we are either disliked, or—let's not beat around the bush—labeled "a bitch." The more a woman succeeds, the worse the vitriol seems to get. It's not just her competence that's called into question; it's her very character. Look at the 2008 election campaign, in which two women ran for high office. Hillary Clinton and Sarah Palin were neatly placed on a continuum that ran from smart and cold to stupid and pretty. No one would ever say those things about a man. All too often, the very fear of this kind of abuse is enough to make women pull too far back and become overly deferential.

Back at the Yale School of Management, Victoria Brescoll has tested the thesis that the more senior a woman is, the more she makes a conscious effort to play down her volubility. It's the reverse of how most men would handle their power. She conducted two experiments on a group of men and women.

First, she asked 206 participants, both men and women, to imagine themselves as either the most senior figure or the most junior figure in a meeting. Then she asked them how much they'd talk at that meeting as their imagined character. The men in the group who'd imagined themselves as a powerful figure reported that they would talk more than the men who'd picked a junior position. But the women who'd selected high-powered roles said they would talk the same amount as the lower-ranking women. Asked why, they said they didn't want to be disliked, or seem out of line or too controlling. Were the women inventing these fears, or were they realistic?

In Brescoll's next experiment, both male and female participants rated a hypothetical woman CEO who talked more than other people. The result: Both sexes viewed this made-up woman as significantly less competent and less suited to leadership than a male CEO who talked for the same amount of time. When the fictitious female CEO was described as talking less than others, her perceived competency shot up.

Not only do we dislike women who talk a lot, we actively expect men to take the floor and dominate conversations; we punish them if they don't. And remember, Brescoll's female participants were just as prejudiced against women as the men.

It's Lonely at the Top

Even women at the pinnacle, women who are tough and normally don't like to complain about discrimination, say they still feel unspoken waves of prejudice in their everyday lives. Linda Hudson, who spent the last four years as president and CEO of BAE Systems, the U.S. arm of the global defense contracting giant, has been a leader in the industry for decades. Still, she is a pioneer, and she told us, "I still think the environment is such that even in the position I am now, everyone's first impression is that I'm not qualified to do the job." "Really?" we asked, taken aback. "Yes," she insisted, explaining the essential corporate difference for men and women. "When a man walks into a room, they're assumed to be competent until they prove otherwise." For women, she says, it's the other way around.

Hudson had just described the reality of "stereotype threat." It's a dry, officious-sounding term, but the experience can be an emotional, confidence-killing loneliness. The term was coined in the mid-1990s by psychologists Claude Steele and Joshua Aronson, who were trying to figure out why African Americans were often still performing less well in college than white students. Since then, hundreds of studies have shown that women, too, underperform in areas like science and math because they are competing in fields in which, stereotypically, females don't do well. Remember that Harvard study from chapter 1, in which women, if asked to name their gender, did worse on the math exam? That's an example of the power of stereotype threat, but the problem is actually much broader. It's essentially a corrosive circle: When we are part of a minority in an institution, and that minority has a well-known stereotype about performance already associated with it, we feel pressure to conform to that type.

And being part of a double minority can become alarmingly

complex. Tanya Coke, a great friend of Claire's since childhood, is an accomplished civil rights lawyer and an African-American woman. She thinks about what she conveys every time she walks into a room that holds strangers. "It's not that I have any apprehension about competing, or that I have a crisis of confidence about my abilities," she said. "I have a consciousness about what people see when I walk into a room. I know they may wonder, until they know me, 'Is this a competent person?' I know I'll be battling assumptions, even if they are unconscious and implicit."

In some ways, *knowing* about negative stereotypes has become a motivator. "I think it makes me all the more intentional about presenting myself forcefully," Coke told us. "I know the challenges I face."

Valerie Jarrett told us much the same thing. "I've never felt it a disadvantage to be a woman, or an African-American woman," she said, pausing, considering her answer. "On the other hand, my parents definitely taught me everything would work out—if I worked twice as hard as everyone." She then bursts out in a hearty laugh. "They told me later they didn't really think that advice would work, but figured they should say it, because it was the best they had."

And for all women, and men as well, the legal framework is archaic. The United States is one of *only three* of the 190 countries in the entire world with no national policy providing a paid maternity leave. New mothers are guaranteed twelve weeks off, but with no salary. That puts us in the same category as women in Swaziland and Papua New Guinea. American exceptionalism sounds nice in theory, but it's often a cold, hard myth for working women.

The latest Global Gender Gap Report compiled by the World Economic Forum ranks the United States not at the top or even in the top ten nations in the world, based on a broad array of measures of equality for women. The United States is twenty-third, just behind Burundi. And in terms of women's political empowerment, by their measure, the United States is a miserable sixtieth. We are first in terms of educational attainment, but sixty-seventh in terms of gender pay equality, placing us just after Yemen. That is a sobering gap.

We aren't laying any of this out as an excuse for not trying new challenges, because the fact is we have to deal with the world as it is even as we try to change it. Still, ignoring centuries of tradition would be shortsighted, to say the least. Understanding the challenge, or the stereotype threat that we face, as Tanya Coke pointed out, can motivate us to battle it.

Confidence and Mirrors

We can't discuss women's confidence and ignore the image in the mirror. We have an extremely tough and limiting relationship with what we see in there. As Marie Wilson points out, we don't know how to use the mirror as a tool for hope or empowerment, since those future senators never make an appearance. At every age, physical appearance plays a disproportionate role in building a woman's self-confidence. We are much quicker to criticize our appearance than men are to criticize theirs. The data is devastating. One international study shows 90 percent of all women want to change at least one aspect of their physical appearance. Eighty-one percent of ten-year-old girls are afraid of being fat. And only 2 percent of us actually think we are beautiful.

We don't know which comes first, attractiveness or confidence. Do attractive people feel more confident? Or do confident people perhaps feel they are more attractive than they really are? What we do know is that there is evidence that women are indeed judged more harshly at work and in life on our physical appearance than men are judged.

Take obesity: The professional penalties for men and women are very different. Christy Glass at Utah State University has studied obesity in men and women, looking particularly at its relationship to education levels. She found that overweight girls were much less likely than other girls to join clubs, get picked for sports teams, and be included in social groups. Teachers even have lower academic expectations for heavyset girls. But the same is not true for overweight boys: they are still included in sports teams, they still date, and are still present in all the important social groups. Obese boys go on to college just as much as other boys do, but obese girls are less likely to go to college than other girls. It sets up heavy girls for a long-term struggle. "Women who don't meet the beauty standard—the social beauty standards—lack critical social resources," says Glass. "They're denied the network ties and people expect less of them." Overweight men can even benefit from the Tony Soprano effect. They can be seen as powerful, savvy, competitive, and intelligent. Yet if a woman is overweight, her size is seen as a negative reflection not just of her physical attractiveness but also of her intellectual capability. She is deemed less organized, less competent, and lacking in self-control.

Women's looks are complicated on all fronts. When Marissa Mayer, the CEO of Yahoo, did a spread in *Vogue* in 2013, critics suggested that she shouldn't be taking time out from her day job to play dress up for a fashion magazine. It seems unfair; however a woman looks, she can't seem to catch a break.

We aren't helping ourselves, either. Our own obsession with our physical appearance drains our confidence. Barbara Tannenbaum is a public speaking coach and beloved professor at Brown University whose course "Persuasive Communication" is so popular it's always standing room only. Video is an important tool in her class. But, with most women, she's found, she can use it only in carefully controlled settings. Tannenbaum regularly videos the class's public speaking exercises so that the students can look back and critique the content of their own performances—did they project enough, make eye contact, engage with the audience, those kinds of things. But initially women can't see any of that, she says, because they are so focused on how they look. "I look too fat, ugly, my hair's a mess . . . it's a huge issue," Tannenbaum says. "I literally have to sit with them, reviewing the video and stopping the self-criticism. You know how little babies need mittens to stop scratching themselves sometimes? I have to be their mittens." And the men in her class? we ask. They may occasionally say they don't like their sweater or that their hair needs a cut, but it'll be a fleeting comment and it doesn't stop them from seeing the important things.

Self-Inflicted Confidence Wounds

Our genetics, our schooling, our upbringing, our society, our looks—these are all factors that affect our confidence. It would be easy to simply shrug our shoulders and blame all those obstacles when we stop short and don't reach for the goals we want. Easy, but misguided. Because we are getting in our own way, too. There are things we do to ourselves, as adults, that kill our confidence. Things that were perhaps inculcated but that we are quite capable of changing.

Look at some of the unhelpful traits women tend to bring into the workforce. We can be exquisitely thin-skinned about our relationships with others and what people think about us. Unlike our male colleagues, women often would rather be liked than respected, which makes it harder for us to shoulder those tough workplace negotiations. The psychic risk of making someone irritated with us is just too great. Claire admits that being liked is essential to her, and it's a need she finds almost impossible to overcome. "I'm not even sure why I need for people to think I'm a nice person or when that started. But any suggestion that bosses or colleagues or even friends might be mad or disappointed starts hours of worry. Lately, as a result of our research, I have come to see that worrying about those things is the opposite of confident."

(At the same time, as Sheryl Sandberg lays out in *Lean In*, studies show that being likeable is also critical for success for both genders, and more so for women, who are actually *expected* to bring likeability to the table. It's a double, in fact a triple bind, if ever there was one. A focus on being liked can kill confidence, and yet, being liked does matter. On top of that, working to be liked may keep us from employing more aggressive strategies that would get us ahead.)

What, exactly, is the cost of our need to be liked? Try $5,000, to start. Finally, you've graduated from college, you even have your master's degree, and that prestigious multinational corporation offers you a job. The salary isn't great. But, hey, you're young, just starting out. After all, you have no experience and aren't you lucky to have a job offer at all? You wouldn't want to make anyone mad by asking for more money. You may laugh, but you know it rings true. That is the soundtrack of too many young women who are just starting out. But it's not how the guy who'll be sitting in the adjacent cubicle thought about it. Which is why he's apt to be making $5,000 more than you.

A study of recent graduates by Rutgers University confirmed the studies we mentioned in chapter 1. It found that's the average pay gap between young men and young women in the first five years after college, and it increases over the years because women don't ask for more money.

If we can be so fragile about the prospect of even mildly annoying someone, it's not surprising we have such a horror of being criticized. Not surprising, but really limiting. If you aren't prepared to be criticized, chances are you'll shy away from suggesting bold ideas, or sticking your neck out in any way. Think of that ad associate Rebecca, fighting back tears in her cubicle just because her boss was critical of a piece of work.

We aren't immune to this weakness, either. Katty realized just how badly she handled criticism when she started getting a lot of it, online, very publicly. It's almost a job requirement for journalists to tweet now, and when Katty set up her Twitter account she was appalled by the responses from people who followed her. "On Twitter people either seem to love me or hate me, and when they hate me they really, really hate me. I get nonstop abuse from people who say I'm an idiot who knows nothing about American politics and should just go back to Britain. One person even said he hoped I'd die." Initially, it was really upsetting, and was almost enough to make her give up. "But, somehow, I got used to it. Perhaps because there was just so much of it, I developed a thicker skin. Now, I find it almost funny. The tweet that said 'I don't give a rat's anus what an uptight pinko Brit chick thinks,' is a particular favorite! One thing about social media is that you quickly learn that you can't please everybody all the time."

Another unhelpful habit most of us have is overthinking. Women spend far too much time undermining themselves with tortured cycles of useless self-recrimination. It is the opposite of taking action, that cornerstone of confidence. There is a formal word for it: ruminating. We do a lot more ruminating than men, and we have to get out of our heads if we want to build confidence.

Susan Nolen-Hoeksema, a psychologist at Yale, spent decades elaborating on the dangers of excessive rumination. Her studies illustrate that women have an instinct to dwell on problems rather than solutions: to spin and spin on why they did a certain thing, how well or (more often) how poorly they did it, and what everyone else was thinking about it. Our intensive capacity for brooding, she maintained, can put us at risk of anxiety and depression. "Over the past four decades, women have experienced unprecedented growth in independence and opportunities," Nolen-Hoeksema wrote in her book *Women Who Think Too Much*. "We have many reasons to be happy and confident. Yet when there is any pause in our daily activities many of us are flooded with worries, thoughts and emotions that swirl out of control, sucking our emotions and energy down, down, down. We are suffering from an epidemic of overthinking."

Before she died in 2013, Nolen-Hoeksema tied rumination to the fact that women are naturally or sociologically more inclined than men to put greater weight on our emotional connections. Of course, our attention to relationships is also one of our greatest strengths. It is what makes women so satisfying as friends. But we are undermining that positive attribute when we spin our emotional wheels too fast. Managers say this tendency for women to overthink is a real hurdle. From her perch running BAE Systems, it frustrated Linda Hudson to no end. Over the years, she's managed thousands of young men and women and has found the same phenomenon Mike Thibault, the Mystics' coach, sees: "Men tend to let things go, slide off their backs. Women tend to be more self-reflective: 'What did I do wrong?' as opposed to thinking it's just a bad set of circumstances and so let's move on."

This is not just a career issue. Unfortunately our propensity to ruminate is not selective. We do it just as much in our personal lives as in our professional lives. How often have you mentally picked away at relationships with friends or a partner, undermining something that was actually perfectly solid? Or spent too many hours second-guessing a decision as simple as whether to get a new haircut?

It was another diagnosis we found all too familiar. While we were writing this very chapter, Katty got into a spiral of self-recrimination. Something had gone slightly wrong at work, there was a new boss, and she was sure she'd disappointed him by saying no to a weekend shift. She came home and spent not just hours, but days, agonizing over it. "I lay awake more than one night thinking, 'I shouldn't have done that. I made the wrong call. That was so stupid of me.' I knew it was minor and would blow over—I didn't even know if he'd noticed, and anyway I could think of fairly good reasons for making the call on that story that I did, but I couldn't stop the tape. It drove me mad."

Have you ever noticed that women tend to be really good at taking the blame for things gone bad, and crediting fate, or other people, or anything but themselves, for successes? Perhaps you've also noticed that men do the opposite. The stories we tell ourselves about the roots of our success and our failure are the foundation of self-assurance.

Dave Dunning, the Cornell University psychologist, gave us an example that perfectly illustrates how the instinct to overpersonalize setbacks undermines women's confidence. At some point in Cornell's math PhD program, he said, the course inevitably gets tough. It is a math PhD program, after all. What Dunning notices is that the men in the course recognize the hurdle for what it is, and they respond to their lower grades by saying, "Wow, this is a tough course." That's what's known as external attribution, and it's usually a healthy sign of resilience. The women in the course tend to respond differently. When the course gets hard for them, their reaction is, "You see, I knew I wasn't good enough." That's internal attribution, and with failure, it can be debilitating. The story becomes one about their intelligence, not about the course itself, or even how hard they worked, says Dunning.

Victoria Brescoll sees the flip side of this at work in the difference between male and female students hunting for their first job. When a young man applies for a position and doesn't get it, she says, his reaction is to blame the process: "They didn't review my application fairly," or "This is a really tough season for job hunting." The woman's automatic reaction is personal: "Oh, no, they found me out, I wasn't up to it." In both these cases, who's more likely to try again? The men Dunning and Brescoll observe have found a great way of dealing with setbacks. It may be emotionally unrealistic—it may be nothing more than denial—but next time there's a challenge that is just beyond their reach, they will be in a far stronger psychological position to go for it. The women will tell themselves there's no point in trying because their past failure proves they themselves aren't good enough. The men will shrug that failure off as an inevitable consequence of external forces, which have nothing to do with their ability. The result is that their confidence remains intact.

But, of all the warped things that women do to themselves to undermine their confidence, we found the pursuit of perfection to be the most crippling. If perfection is your standard, of course you will never be fully confident, because the bar is always impossibly high, and you will inevitably and routinely feel inadequate.

Moreover, perfectionism keeps us from action. We don't answer questions until we are totally sure of the answer, we don't submit a report until we've line edited it ad nauseam, and we don't sign up for that triathlon unless we know we are faster and fitter than is required. We watch our male colleagues lean in, while we hold back until we believe we're perfectly ready and perfectly qualified.

It didn't surprise us to learn that plenty of studies show that this is a largely female issue and that we manage to extend the perfectionist disease to our entire lives. We obsess about our performance at home, at school, at work, on holiday, and even at yoga class. We obsess as mothers, as wives, as cooks, as sisters, as friends, and as athletes. The irony is that perfectionism actually inhibits achievement. Bob Sullivan and Hugh Thompson, authors of *The Plateau Effect*, call it the "enemy of the good," leading to piles of useless, unfinished work, and hours of wasted time, because, in the pursuit of it, we put off difficult tasks waiting to be perfectly ready before we start.

Even Professor Brescoll, well aware of the research, still has a hard time controlling her penchant for perfection. Academics are judged on how much they publish and in which prestigious peer-reviewed journals their work appears. Brescoll confesses that she often takes much longer than her male colleagues to submit a paper for consideration, determined to get things exactly right before pressing the send button. Sometimes she'll aim lower, sending the paper to a lesser journal. "I need to be much more certain before I'll take a risk of submitting my work. My male colleagues just send stuff like crazy, they take a chance. Sometimes it works, and sometimes it doesn't," Brescoll says, "but in the end, their strategy is more effective, due to the sheer volume."

With so many self-imposed obstacles for growing confidence, it's a wonder we have any at all. But transformation is often so simple. Brescoll has learned, for example, that if she just puts her work out there, without obsessive thought, things happen. Either it's accepted, or, if not, she's learned to value the feedback that comes with rejection. It lets her make corrections, and then try again. That's the cycle that breeds excellence and mastery, allows us to stretch our limits, and creates self-assurance.

It was at this point, just when we'd come to the encouraging conclusion that many of the reasons women have less confidence is actually due to factors we can control, and therefore diminish, that a friend of ours sent us in an unexpected direction. She's a doctor, and suggested we look at gender differences in brain biology and operation. We quietly groaned. We hadn't found overt differences between men and women in our investigation of some of the primary genetic influences on confidence. The gene variations we'd examined, which influence transmitters like serotonin and oxytocin, are dispersed in equal measure between genders, we were told. Frankly, we were happy to stop there. The idea that male and female brains might not function in exactly the same way seemed too complicated, and potentially loaded. But it seemed undeniable, too. Of course our brains work differently, as we thought about it. We'd just spent months investigating how. Women ruminate; we stew, we want to get everything right, and we want to please and be liked. The two of us realized we needed to broaden our look at the science behind confidence. We'd mapped out some key behavioral differences that affect confidence, but what is actually going on inside our heads that might play a role?

It Matters Where the Matter Is

The very suggestion that male and female brains might be built and function in unique ways has long been a taboo subject, largely because among women, it was generally thought that any difference would be used against us. That's because for decades, for centuries actually, differences (real or imagined) *were* used against us. Based on no evidence at all, women were deemed to have an unbearable lightness of thinking. And it's still a subject that can produce tremors. As recently as 2006, Larry Summers, then the president of Harvard, found himself embroiled in controversy after suggesting, based on his reading of research, that there may be innate differences between men and women when it comes to achievement in science. Eventually, the lingering anger over those remarks led to his resignation.

So let's clear the air: Male and female brains are vastly more alike than they are different. You couldn't stare at a scan of two random brains, and clearly identify male versus female. In terms of intellectual output, the differences are negligible. That doesn't mean there are no differences though—there are, and some of those differences in structure and matter and chemistry may encourage unique patterns of thinking and behavior, patterns that can clearly affect confidence.

In terms of sheer size, men do have women beat. Their brains are indeed larger and heavier relative to their body size. Does that mean male brains are better? No. IQ tests are basically equal for the two sexes, though in some measures men tend to score higher on math and spatial skills, and women routinely outpace men on language arts.

One Harvard study found distinct differences in the distribution of our brain matter, which suggests vastly different methods of processing information. Women tend to have the bulk of their brain cell matter in the frontal cortex, the home to reasoning, and some in the limbic cortex, an emotional center. Men have less than half of all their brain cell matter in their frontal cortex. Theirs tends to be spread throughout the brain.

There are actually two types of brain matter, gray and white. Men have more of the gray stuff, useful for isolated problems, and women have more white matter, which is critical for integrating information. It's almost as though evolution designed our brains to reach equally complicated destinations on completely different roads, neurologist Fernando Miranda, an expert on learning disorders, told us.

For Jay Lombard, the Genomind neurologist, the most compelling evidence that there are material differences in male and female brains comes from a series of studies involving DTI (diffusion tensor imaging) technology. These scans are seen as increasingly useful for studying function, because they can map the connectivity of our brains. Essentially, DTIs can study the integrity of our white matter, the essential tissue that fosters connections. A handful of studies have found that women tend to have better functioning white matter, and in some important places, like the *corpus callosum*, the central highway between left and right brain. Dr. Lombard thinks it might explain why women work with both sides of their brains more easily. Miranda thinks white matter might be why women are often faster at making multiple mental connections and are more adept at broad reasoning. "The work is in the early stages," says Lombard, "but there is no question that there are significant differences."

One leading psychiatrist, Dr. Daniel Amen, recently combed through 46,000 SPECT brain scans, which measure blood flow and activity patterns. He, too, found some notable differences between genders, and wrote a book called *Unleashing the Power of the Female Brain*. Amen, we should note, has made a successful and lucrative career out of popularizing theories about the brain. He's the author of thirty books, including several best sellers, and is a regular guest on *The Dr. Oz Show*. Some critics think he overstates his findings. We carefully compared his work with other studies, talked with a number of experts about them, and discovered that even his detractors believe his research is important. Some of it was remarkably relevant to our investigation.

What Amen found is that female brains are more active, in almost all areas, than male brains—and especially in those two areas we mentioned, the prefrontal cortex and the limbic cortex. One study suggests that women have 30 percent more neurons firing at any given time than men. "The activity in these regions probably indicates female strengths, including empathy, intuition, collaboration, self-control, and appropriate worry," Amen told us, "but women are also more vulnerable to anxiety, depression, insomnia, pain, and being unable to turn off their thoughts." In other words, his scans seem to be showing, physically, all of our overthinking, our ruminating, in process. "When the prefrontal cortex works too hard, as it often does in women," explains Amen, "it's like the parking brake is always on and you can get stuck on certain thoughts or behaviors, such as worrying or holding grudges." Dr. Amen believes his study is evidence that women *are* often thinking more than men. That can be a plus. It's why we are better at multitasking, he says. But it can also lead to that snowballing of negative thinking and anxiety. "It's useful in small doses," says Amen, "but then it becomes worry and stress to the point where you can't rest." A confidence killer.

As we looked at other structural differences, we learned more about those primitive fear centers, the mysterious amygdalae. First, we all have two of them, not one, and they do rather different things. One is associated with taking external action as a result of negative emotions, and the other with using thought processes and memory in response to stress. And, yes, you guessed it, men rely more on the amygdala that deals with action, while women tend to activate the memory/emotion amygdala more easily. It's a reflection, in our brain structure, of the notion that men seem to respond to challenging or threatening situations with action, while women favor internal mechanisms.

Add to that a new study from McGill University in Montreal, which shows that women produce 52 percent less serotonin in their brains than men do—remember that's the critical hormone that helps keep our anxiety and amygdala under control. We started digging into women and serotonin levels and we learned a lot we weren't eager to hear. While girls aren't born with the short strand serotonin variation any more often than boys are, it turns out that girls and women respond differently to that variant. When women get the short straw, or the short strands, they are more likely to be prone to anxious behavior than men are. Likewise, some studies show the same result for women and the COMT variant that controls dopamine. When we get the "worrier" variety, we are more likely to be truly anxious.

We were coming to see that a woman's thinking machine is far from perfect for generating self-assurance, when we uncovered one more physical difference that we found particularly loathsome. There's a little part of our brains called the cingulate gyrus. It helps us weigh options and recognize errors—some people call it the worrywart center. And, of course, it's larger in women.

Terrific, we thought. How wonderful to know we have a *special department* for that useful habit. How maddening to be proven to be exactly the picky worriers our husbands sometimes accuse us of being. There are, we should emphasize, many positive aspects of our brain's behavior. In evolutionary terms, we needed to be the cautious worriers, always scanning the horizons for threats. We are built superbly for that. Today, however, those particular tools may not be as useful or enjoyable.

A crucial difference we have that *is* a clear advantage in our modern lives—women tend to use both hemispheres of our brain more regularly than do men, combining the left side, home of mathematical and logical skills, and the right side, where the artistic and emotional skills reside. That's the science of female multitasking abilities. Laura-Ann Petitto told us that bilateral use of the brain is more effective, and, actually, more cognitively advanced. That news we liked (though we resolved not to toss out the theory at home).

Some of these life-determining brain differences begin before we are born. A study by Israeli researchers looking via ultrasound at male and female brains in the womb found differences as early as twenty-six weeks. In the largest ongoing study of brain development in children, the National Institutes of Health has documented that, by age eleven, there's a large gulf not just between the way boys and girls think but also in their actual capabilities. Essentially, young boys are well behind their female peers in both language ability and emotional processing, but girls are almost as far behind the boys in spatial ability. The anatomical difference in capabilities is usually resolved by age eighteen, but that gap, if misunderstood, can easily reinforce stereotypes at a critical learning age. You can imagine why, at sixteen, a girl may conclude she's bad at math, or a boy may declare that he will never get Shakespeare. And yet, if they just waited for their hormones to settle by their late teens or early twenties, the necessary brain functioning for both math and Shakespeare would be online in both boys and girls.

The Risky Business of Testosterone

We'd figured testosterone and estrogen were essentially the celebrity architects of gender difference. A bit showy, attention-seeking, and overexposed, they command enormous respect, and for good reason. Everyone knows they are the forces behind many of the basic, overt differences between men and women.

We didn't think they would be significantly involved in the detail work though—the intricate creation of something as complicated as confidence. Certainly, we thought, our confidence differences can't spring from such an elemental source. But those hormones seem to be considerable players here, too. Testosterone, especially, helps to fuel what often looks like that classic male confidence. Men have, after puberty, about ten times more testosterone pumping through their systems, and it affects everything from speed to strength to muscle size to competitive instinct. Testosterone is the hormone that encourages a focus on winning the game and demonstrating power, instead of connecting and cooperating with others.

Testosterone is also heavily correlated with risk taking. A number of recent studies have tied high testosterone to a willingness to ignore traditional cues about risk. We detailed a fascinating one in our last book, and each time we mention it in a speech we still draw nervous laughs. Scientists from Cambridge University followed seventeen male traders at the London stock exchange, all high rollers, for a week. They earned good salaries; many collected bonuses as high as $5 million. Researchers measured their testosterone levels using saliva samples, at the start and end of the day. On days when traders began with higher levels of testosterone, they made riskier trades. And, when that paid off, their testosterone levels didn't just rise—they spiked. One trader who doubled his take saw his testosterone level almost double along with it. Testosterone fuels risk taking, and winning creates more testosterone. The dynamic, called the "winner effect," can be dangerous. Animals can become so aggressive and overconfident after winning that they take fatal risks, like standing in open ground, prompting other animals to attack them.

Higher testosterone levels are linked to feeling powerful. When women are told to adopt classically male seating poses—taking up more space, spreading their legs and arms—our testosterone levels rise. Power poses are a popular tool for communications classes. Barbara Tannenbaum uses them in all her presentations. She'll start the discussion by asking the men in the audience to sit like women and the women to sit like men. After years of doing this, she's made two observations: First, the exercise always raises a laugh. Second, no one ever asks her what she means; they just know. Often the men will say it's really uncomfortable trying to sit with their legs crossed and their bodies somehow shrinking inward. (Try doing it in Spanx and heels, she quips back.) The women, though, seem liberated by the unfamiliar pose. One day Tannenbaum did the exercise in a high school in India, and, as she spread her knees and lolled back in her chair, one young woman blurted out, "I feel like a king when I sit like this!" *The confidence of kings*—that's what we'd like to give all our girls.

There's a downside to testosterone, to be sure. This egocentric hormone is also associated with an inability to see others' points of view. When you have a lot of testosterone coursing through your body, you're less interested in connecting and cooperating. That's not great for business in a modern world, which relies so heavily on communicating with others. The finding of one experiment suggests women can fall victim to testosterone's perils as well. There were thirty-four women in the study, who were split into two groups and told to work in pairs to examine the clarity of some computer images. Some of the images were obviously clear; others weren't. If the women disagreed on which image was sharper, they had to collaborate and agree upon a final answer with their partner. The first group of women was given a testosterone supplement; those in the second weren't. Go figure—the women on testosterone were both less able to collaborate *and* wrong more often.

The main hormonal driver for women is, of course, estrogen, which promotes very different instincts from testosterone. Estrogen encourages bonding and connection; it supports the part of our brain that involves social skills and observations. It is part of what drives women to avoid conflict and risk and so it might hinder confident action at times.

But there are substantial performance upsides to estrogen as well. A testosterone-laced decision isn't always better. The big risks that ensue often lead to spectacular failure, as the world economy has witnessed. Indeed, several studies of the strategies of female hedge fund managers show that taking the longer view and trading less can pay off. One study in particular got us thinking about how to harness our natural strengths. Researchers found that the investments run by female hedge fund managers did three times as well as those run by male managers over the past five years. And the women lost significantly less money than men in that disastrous year, 2008.

So, what are the implications of all this brain research in the search for greater confidence? Well, there's good news and bad. Women's hormonal tendency to avoid risk and conflict can lead to too much caution. Our laborious brain processing mechanism can become a maelstrom of overthinking and indecision.

However, the science also suggests that women have the ability to be hugely accomplished and successful. Our brain structure means we like to get stuff right, make good judgments, and hold bad impulses to a minimum. Our biological investment in emotions makes us good at perceiving problems, at understanding the issues of others, and at moving toward reconciliations and solutions. And our highly integrated brain means we can take in large amounts of data and process it quickly. Stepping back, we realized how much of this tracks with what we see in our behavior. We'd just discovered the action going on backstage.

The essential question still to be answered about all of these gender variations, however, is whether male and female brains are *programmed* to develop this way. Perhaps some of these physical differences are a result of the way we are raised. Do women end up with connective white matter because we grow up using it more? Do we have more of it because *centuries* of women have grown up using it more? Scientists aren't close to a global explanation, although, remember, there is growing evidence in all of the work on brain plasticity that our brains certainly can change in response to the environment over the course of our lives. We uncovered one startling study that found testosterone levels in men decline when they spend more time with their children. (The implications are well worth further study.)

Whatever the cause-and-effect relationship, we found this link between brain structure and behavior incredibly useful as a visualization exercise. For us, it was a snapshot of the internal mechanism that may be at work when we see common behavioral patterns. We started to take our ruminating habits more seriously, and yet we were able to give ourselves a psychological break for them, because we knew that larger forces were probably involved.

We were also able to face the fact that men and women act in curiously different ways with somewhat more equanimity, and a good bit less emotion. We thought, for example, that all of this research might explain why the men we know seem to be able to shrug off disagreements with people so much more easily than we do. We both see it in our own husbands. They have a fight with a friend and, poof, a few minutes later the tension has disappeared. Where we waste weeks with worrying and hurt feelings, they come out of a blazing argument and a day or two later can barely remember it happened.

We came away from our look at the brain biology feeling surprisingly empowered. The way we often think and behave isn't wrong, but rather, understandable. At least, that's how we decided to regard it—using self-compassion. All we need to do is get our natural instincts to work more in our favor.

We also felt thoroughly versed in our challenges. We had a gratifying sense of mastery, you might even say, of the societal issues, the science, and the behavioral patterns affecting confidence. Fully armed, we were ready to move on to confidence creation.

5

THE NEW NURTURE

Jane's mother walked her to school for the first day of class; around Jane's neck, tied to a string, was the key to their front door. Jane knew that at the end of the day she would have to retrace those steps all alone, let herself into the apartment, and wait an hour and a half before her big sister got back home. She was terrified and remembers crying at the prospect of something going wrong. But she made it, and after the first day she did that solo journey again and again. She was four and a half years old.

"By six, I was in Girl Scouts and I knew I'd be a leader because I'd already done the tough stuff," she says now. "My mother didn't show me any fear, she just said she knew I could do it and find the way home by myself. It was doing hundreds of those small things that built my confidence as an adult. You're not born with it. You build and you build. I built that confidence myself."

When we first heard that story, we were taken aback, to say

the least. Alone! At four and a half! Anything could have happened! What kind of parenting is that? Maybe the right kind. We were coming to understand that the acquisition of real confidence requires a radically new kind of nurturing, of ourselves and of our children. It demands more than just praise and love and hugs and making things easy on them (and on ourselves). It demands more than just the pursuit of good grades and perfection. None of that is working, and certainly not for women and girls. The conclusion of Jane's story suggests that her mother may have done her a big favor.

Jane Wurwand, all grown up now, is the founder of the skin care company Dermalogica. She's a Brit who has made it big in America, with a little help and a lot of determination, and now lives in Los Angeles with her husband and two teenage daughters. Twenty years ago, her confidence enabled her to take a substantial risk. Turned down for multiple loans, she put her own life's savings of $14,000 into the company that would become a global brand and multimillion-dollar business, with more than fifty offices around the globe. As Wurwand says, with a smile, "Not bad for a girl with only a beauty school diploma from a small town in England."

Little Christine was presented with much the same challenge and even more responsibility. At the age of four, she was expected to babysit for her younger brothers. If her parents had plans out in Le Havre, where they lived in northern France, they'd simply say, "We're going out, look after the others." She tells the story of one particular evening when her parents went to a concert, leaving her in charge. They said they'd be back at eleven. The appointed hour came and went, and still no parents. She turned on all the lights in the house to keep from getting scared and went upstairs to check on the baby. Finally her parents got home and found her curled up, reading a storybook in her brother's bedroom. There was no reason for panic, the parents felt; they'd simply decided to go out for a bit after the concert, and why not? All four-year-old Christine said was, "Well, you're a bit late."

Today, IMF Managing Director Christine Lagarde laughs at the absurdity of her parents leaving such a young girl to babysit. But there is little question that parents of a different generation, and perhaps from different cultures, have offered a hands-off approach that in many ways served kids better than our modern American meddling. Lagarde says, with total conviction, that in doing so her parents were creating a cycle of trial, responsibility, and success that helped build the confidence she wields today on a world stage. It didn't end with the babysitting. When Christine was sixteen, her mother dropped her by the side of the highway to hitchhike down to Lyon, six hours away, to visit some friends. At twenty, Christine was sent off to America alone, armed only with a plane ticket and Greyhound bus fare. "She equipped me with that sort of sense of 'you can do it.'" And do it she did.

We are not advocating a new home-alone policy for toddlers, and truth be told, we ourselves don't embrace the notion of leaving four-year-olds in charge of younger siblings, as much confidence as it might create. But you get the point. We share these eyebrow-raising stories to give you a jolt as we explain the new nurture. Because nurture, in order to create enduring confidence, needs to toughen up, to shake off that warm and fuzzy image.

Think of it this way: Much of what parents have been told to emphasize for the past twenty years, based on the self-esteem movement, is misguided, and it's generated a glut of flimsy self-esteem and flimsy confidence. Children have been rewarded for anything and everything, instead of for genuine accomplishment. Girls and women express more self-doubt than boys and men do, but modern parenting has created hollow confidence for both genders, as it often gives kids little responsibility, matched with a lot of praise and prizes. They're deprived of adversity and the chance to fail. It's the opposite of the type of parenting that accomplished women such as Lagarde and Wurwand received.

In some ways, false confidence is even more damaging than hollow self-esteem, since confidence is about ability and mastery. If you believe you can do something, and you truly can't, and you are no longer an overprotected child, that clash with reality will be painful. This is not akin to wielding the bit of overconfidence Cameron Anderson says can be useful. We're talking about a yawning gap that can cause real problems.

Scores of employers report that many of today's college graduates, for example, seem to think they can run the world straight out of college and that they *deserve* every job they apply for and every perk they can get their hands on. Dig a little deeper, and you'll find that their confidence is actually very fragile, because it has so little foundation in experience or reality. They may sound like know-it-alls, but push today's youth, and they crumble fast. And their parents bear a lot of responsibility.

"In the past, you would gain confidence by trial and error, and over time you begin to learn, 'I'm generally right, I can do things,'" says psychologist Richard Petty. Today, children get praise from parents who think they are boosting self-confidence but who are actually just indulging kids who have done little to deserve it because those very same parents have prevented their precious offspring from losing and failing and risking. At some point, usually after the children leave the confines of that overly protective nest and encounter the big, cold world of work, reality intervenes. "Things get objective, and people tell you when you've made a mistake. It isn't all just roses," says Petty.

Hard Knocks and *Gaman*

So, what's the magic formula? For once, we found surprising clarity and consensus. Confidence, at least the part that's not in our genes, requires hard work, substantial risk, determined persistence, and sometimes bitter failure. Building it demands regular exposure to all of these things. You don't get to experience how far you can go in life—at work and everywhere else—without pushing yourself, and, equally important, without being pushed along by others. Gaining confidence means getting outside your comfort zone, experiencing setbacks, and, with determination, picking yourself up again.

Maybe we've all grown a bit soft in the postwar, baby boomer years, and it's time to toughen up, to develop a thicker skin and more resilience, because our investigation showed us clearly that enduring a few hard knocks, ideally early in life, is the fastest, most effective route to confidence.

Nansook Park, the University of Michigan psychologist who is an expert on optimism, says that, in general, the proper way to build confidence in children is to offer them graduated exposure to risk. Trauma is not the goal. "They should be introduced to risk taking, but carefully. Don't just drop them in the middle of the lake. Teach them how to do things, and then give them opportunities, and be there when they need guidance. When they succeed, celebrate together, and talk about what worked. And if they fail, talk about what they did well, and the action should be the emphasis, but also what they can learn, and how to make it better the next time."

Failure. There it is. It's the most frightening, and yet most critical partner to confidence. Failure is an inevitable result of risk taking, and it's essential for building resilience. Petty says there's not enough of it. "Just watch *American Idol.* There are kids who can't sing but think they can, probably because everyone always told them, 'Oh, you're great. I like you.'"

We've learned that the secret to success may in fact be failure. By failing a lot, and when we're young, we inoculate ourselves against it and are better equipped to think about the big, bold risks later.

Failure, though, has to be handled in a constructive fashion, Chris Peterson, Park's former colleague at the University of Michigan, explained to us. "One of our students told us about a colleague who was teaching in a tough inner-city high school and bragging about the fact that he'd told his class, 'You're dumber than a bunch of five-year-olds.' And I'm thinking, what a terrible message. You're not only saying you are not doing well, but also you can never do well. A better message is: 'Okay, you made a mistake. You didn't succeed. Here's why. Maybe you can try a different strategy.' That's where confidence comes from."

Of course, part of risk and failure means pushing ourselves and our children in areas in which we're not comfortable. That's novel for many Americans, but in Asia it's the canon of parenting. Asians are all about grit, the hot new term in positive psychology circles for persistence and tolerance for hardship. In Japanese they even have a word for it—*gaman.* Roughly translated it means "keep trying," and it gets plenty of use.

Elaine Chao, the former labor secretary whose family moved to the United States from Taiwan when she was a child, believes the concept of embracing adversity is something the West would do well to learn from the East. "Americans play to their strengths; maybe it's part of their Christianity that says 'God has given you certain talents and you must go and develop them to your full potential.' You talk to an American and they say, 'I'm awful at math, so I'm not going to do math, I'm going to do writing.' But the Chinese proclivities are very different. The parents try to shore up their child's weaknesses. If a child is bad at math, the prevailing wisdom is to focus on improving that subject."

As an immigrant, Chao had a bumpy childhood; one that we might have thought, before our research, would be scarring rather than enriching. She is the oldest of six girls. Her father escaped a small village in China when the communists took over and the family moved to Taiwan, where he won a scholarship to come to America. It was three long years before he could save enough to bring his young family over to join him. During that time, her mother was effectively a single parent and Elaine stepped in to help. "I think birth order is critical. I was the eldest child. When we first came to America, we went through some very tough circumstances," she told us. "My parents were depending on me. My sisters were depending on me. There was no choice but to put on a brave face and get things done."

When Elaine arrived in Queens at the age of eight, speaking not a word of English, she entered third grade at the local school. It was total immersion into a set of unsympathetic classmates. "It was 1961 and America was not as diverse as it is now. It was basically black and white." As a non-English-speaking Chinese girl, she was a tantalizing target to tease. To this day, she remembers the boy who caused her difficulties. "There was a boy named Eli who was the bane of my existence because I didn't understand English, so I couldn't differentiate clearly between 'Elaine' and 'Eli.' I couldn't hear the difference. So whenever his name was called, I would get up and everybody laughed at me."

She couldn't figure out how to fit in at school, and at home, things were tough. There was little money and there were no other relatives; the family lived a very isolated life. As the oldest, Elaine was expected to work hard not just for herself but also to help lift up her five sisters. She credits those days for the confidence she has now. "I'm not sure a nurturing environment is always good. Some adversity, if it doesn't break you, does make you stronger."

This all sounded right, all of the advice about risk and failure and grit and embracing the pain of life in order to learn. With our brains, we believed it. Our hearts, however, refused to cooperate. Despite all the evidence we'd amassed, we struggled, and still do, with actually dealing out the tough-love version of nurture. Often our maternal hearts triumph, and we instinctively, physically, can't stop ourselves from intervening to make the world smooth again for our kids. You might think we parents would be more protective of girls than boys, right? But the evidence suggests that boys, and firstborn boys in particular, also suffer from what Katty's father likes to call "an insufficient lack of neglect," or too much cosseting.

Claire's been forced to embrace the beauty of failure as the mother of two kids who love sports. Given that she had no athletic experience to speak of growing up, and didn't much like to fail, it's been rough. Her son Hugo has become a baseball player, and in the early years, she found each at-bat excruciating. She would experience his emotions more keenly than he felt them, almost unable to watch. And then, ironically, she became her daughter's soccer coach when nobody more qualified could do it. "We were already doing the research for this book, and Della is so athletic; I knew she and the girls needed a team. But it was surprisingly stressful. I'd worry about Della when she got mad at herself for letting a goal in, then I'd worry on behalf of the whole team if they lost. I just couldn't let go of the failures, and I barely saw the successes." Finally, another coach suggested she just embrace the whole experience, losses, struggles, and all. "In team meetings I started pointing out even the small successes each girl had, which they loved, and which helped me, too. By the end of the year, they were ferocious out there." She watched her son put in extra work for months on his batting, and then try out and make the team he wanted to be on. "I'm learning as much about failure and struggle and mastery as the kids are, and I clearly needed to."

Patti Solis Doyle learned that the benefits of risks and failure extend well beyond childhood. Seven years ago she gambled big. She had worked for Hillary Clinton for years, and she said yes to running her presidential campaign. She knew it would be tough, potentially thankless, and maybe disastrous, as is often the case with political campaigns. Moreover, she would be one of a very few female campaign managers, and the first Hispanic woman to run a presidential bid. No small stakes.

A year later, with poll numbers less than ideal, she was fired. She took the dismissal hard and felt badly bruised. For months, she was convinced she'd never work again, but she slowly started to embrace what had happened.

"In retrospect, I'm so glad I took that risk," she told us. And then she laughs, shaking her head at the memories. "Now—there's no way I would have said that right after I lost my job, not at all." Solis Doyle echoes Elaine Chao: "I've come to see that what doesn't kill you makes you stronger. I've learned so much. I've learned how to deal with negative stories; I've learned that we can lose and move on."

She's since mended political fences and has launched a financial start-up that buys debt from state governments. It's not an area in which she felt she had any previous experience, but she decided to try it, she says, because she knew she could handle failure as well as success. She recently sold the company to a huge corporation for a hefty profit.

The Talent Myth

The starting point for risk, failure, perseverance, and, ultimately, confidence, is a way of thinking, one brilliantly defined by Stanford psychology professor Carol Dweck as a "growth mind-set." Make an effort to read anything she's written. She's found that the most successful and fulfilled people in life always believe they can improve, that they can still learn things. Let's go back to the example of how women and men approach their math skills. Most women think their abilities are fixed, Dweck told us. They're either good at math or bad at math. The same goes for a host of other challenges that women tend to take on less often than men do: leadership, entrepreneurship, public speaking, asking for raises, financial investment, even parking the car. Many women think, in these areas, that their talents are determined, finite, and immutable. Men, says Dweck, think that they can learn almost anything.

Confidence requires a growth mind-set because *believing* that skills can be learned leads to *doing* new things. It encourages risk, and it supports resilience when we fail. Dweck has found that a growth mind-set especially correlates with higher levels of confidence for adolescent girls.

The growth mind-set can help us recast failures as critical learning experiences. Katty immediately saw that Dweck's way of thinking had been a missing ingredient for her. "I've always assumed there were things I was good at (languages, child rearing) and things I wasn't good at (business, athletics, management, anything involving a ball, musical instruments, computers—the list is long). The one that causes me most regret is being entrepreneurial. I've always admired people who set up businesses and would have loved to build one myself, but I've never had the confidence to try. It's not that I assume I wouldn't be good at it—I assume I'd be terrible at it, that I'm not a natural businessperson. Reading the news, live, to millions of people every night, I feel quite confident, but setting up a business—total panic. I realize now from all the research we've done that the only way to get the confidence I need would be to give it a try. And Dweck's work allows me to see this as a skill set to develop rather than one I innately have or don't have."

The key to creating a growth mind-set is to start small. Think about what you praise in yourself or your kids. If you praise ability by saying, "You're so smart" or "You're so good at tennis; you're a natural athlete," you are instilling a fixed mind-set. If, however, you say, "You've worked so hard at tennis, especially your backhand," you are encouraging a growth mind-set.

Making a distinction between talent and effort is critical. If we believe that somehow we're given talents at birth that we can't control, then we're unlikely to believe we can really improve on areas in which we're weak. But when success is measured by effort and improvement, then it becomes something we can control, something we can choose to improve upon. It encourages mastery. And it's a part of the Asian *gaman* approach, in fact. Chao admits it can be tough on kids forced to work hard at something at which they will never excel, but it does allow you to take control of your self-belief. Confidence becomes less about what you were born with, and more about what you make of yourself.

For Chao, in the workplace, that means encouraging people, particularly women, to push themselves, to take on tasks that they think are beyond their reach, like leadership. Claiming a top position always seems daunting, but coasting at the same level doesn't increase our confidence. The trick is to recognize that the next level up might be hard, that you might be nervous, but not to let those nerves stop you from acting. "Every leadership position is a stretch," Chao says, speaking of her own experience. "No one ever thinks they're born a leader, that this or that leadership position is perfect for them. It is always a stretch. We should just encourage young women to stretch more."

The Softer Side of Munitions

"There was just always something different about me," said the petite blond woman sitting in front of us, dressed in pastel and grinning broadly. Her office upholstery is soft and floral, but her artwork, we notice on closer inspection, is made from bits of weaponry. Linda Hudson has plenty of experience breaking molds. She's the first woman to head a major defense company. Before becoming CEO of BAE Systems, she was the first female company president in the history of General Dynamics. She was also the first woman vice president of Martin Marietta, and the first female manager at Ford Aerospace. At the University of Florida School of Engineering she was one of only two women in her class. No surprise, then, that in high school Hudson was the first girl to take the engineering drawing course.

She's comfortable with being unique, although Hudson says we should not mistake that for a suggestion that her path has been easy. She's felt more than her share of isolation in the most masculine of masculine industries. After all, as she quipped, with her characteristic bluntness, "We make munitions and tanks and things like that."

Being different is part of the story of every highly successful woman, just by virtue of the fact that there are so few women at the top. We can resent it, let it undermine us or limit us, or we can embrace our uniqueness and choose to wear it as a badge of honor. The earlier you learn to take the risk of standing out, the easier it will be to stand up for yourself in a tense negotiation, demand the high-profile assignment that your male colleague will otherwise grab, or do all the other things that don't fit with the stereotype of a docile, good girl.

Caroline Miller, the author and psychologist who specializes in confidence and optimism, says a willingness to be different is critical to confidence. "It's more than just risk and failure, though those are essential," she says. "Confidence comes from stepping out of your comfort zone and working toward goals that come from *your own values and needs*, goals that aren't determined by society." That realization changed the course of Miller's life. As a young woman, she struggled with bulimia. A top student at Harvard, who then went on to a lucrative job on Wall Street, she kept her secret well-hidden. Finally, in crisis, she got help, and later went public with her first book, *My Name is Caroline*, a hard, honest account of her illness. Soon thereafter, Miller got a master's degree from the University of Pennsylvania's Positive Psychology Center and started a new career.

It's a sense of self, rather than just a sense of achievement, that we can teach early. If Linda Hudson's parents wanted a traditional, girly girl, they didn't get one, and, more critically, they didn't try to create one. She describes herself as a street fighter who grew up preferring basketball with the boys to ballet with the girls. Her favorite subject was math. She says she never felt pressured by her parents to try to be anything other than what she wanted to be. Hudson says she's never been interested in being liked. She wants to be respected.

She thanks her parents for that, too. Her parents were teachers; they didn't have much money, but instilled in their willing daughter the value of learning and the confidence to dream big. More important, when she faced a setback, they were the ones who would send her back into the world.

You don't need to be a roaring Tiger Mom to push your child to work hard and take risks. Hudson's parents did it with love and open minds. As Hudson carved a couple of hours out of her tight schedule to talk to us about her career of firsts, what emerged was a picture more complex than the simple stereotype of a tough woman in a man's world. She is disarmingly frank, both about her successes and her weaknesses. She has no trouble saying that she's competent at what she does and is equally happy to talk about where she needs to improve. ("Listen more, talk less.") She's even open about some personal regrets.

Hudson takes pride in being different, and the one time she succumbed to pressure to fit in, it didn't work out. "I got married right out of college and I would say it was largely because it was expected that I got married. I changed my name, largely because it was expected that I would change my name." After twenty-five years of marriage, she and her husband divorced. She said wistfully, "I'd love to have my name back." However, by then she'd also spent twenty-five years building a career. "I have a professional reputation tied to the name I have now. So, it was just too hard to change it back, and for what point?"

Just as we found with other senior women we interviewed, Hudson's very openness added to her aura of confidence. It occurred to us that genuinely confident women, perhaps genuinely confident people, don't feel that they have to hide anything. They are who they are, warts and all, and if you don't like it, or think it is weak to show vulnerability, too bad for you. These ambitious women have taken a risk in exposing their weaknesses, but it definitely hasn't kept them from succeeding; indeed it may well be part of the reason for their success: They are brave enough to be not only different, but to be themselves.

Suppress the Siren Call of Praise

Think about how terrific it feels when you get a compliment on your work, your clothes, your hair. Often, there's an immediate lift, and then we relive those moments later to rekindle that buzz. It turns out that flattery and praise are as lethal as sugar. A little bit is fine, but much more than that and we're unhealthily addicted. Ohio State University psychologist Jennifer Crocker has discovered that people who base their self-worth and self-confidence on what others think of them don't just pay a mental price; they pay a physical price, too. Crocker's study of six hundred college students showed that those who depended on others for approval—of their appearance, grades, choices, you name it—reported more stress and had higher levels of drug abuse and eating disorders. The students who based their self-esteem and confidence on internal sources, such as being virtuous or having a strong moral code, did better than the others in exams and had lower levels of drug and alcohol abuse. Other studies suggest that men rely less on praise than women do to feel confident.

Confidence that is dependent on other people's praise is a lot more vulnerable than confidence built from our own achievements. Even the most accomplished, beautiful, and celebrated human beings you know don't get a nonstop stream of compliments and positive feedback.

Of course it's unrealistic to think that concert pianists won't compare themselves to their peers, but if your confidence comes solely from your rankings, the press reviews of your last concert, and the adoration of your fans, what happens when those dwindle? Better to have developed the solid, internal confidence that comes from knowing you worked hard to earn a seat in a revered orchestra and play alongside the best in the world.

For the rest of us, we naturally like the satisfaction of good grades, a nice salary, a wonderful email from a boss, but we have to remain connected to our own pleasure at a job well done. When our confidence is based on external measures, the biggest risk is that we won't act. We are more likely to avoid risk if we think we might feel a dip in approval. Chasing permanent praise can lead to self-sabotage. Raising our children to constantly seek our approval, instead of helping them to develop their own code, will be debilitating for them later.

Katty, for example, used to worry that Maya, her eldest daughter, was too much of a people pleaser, too much of a good girl. She was so responsible that everyone in the family relied on her to babysit, cook, help out, and be polite and diligent, and Maya never complained. Sometime in her teens, she blossomed and became much more sure of what she wanted, but it took Katty awhile to see that as a good thing. She developed a healthy, though often infuriating, stubborn streak. "Take this summer, when I was pushing her to start a college application essay that was due mid-October. Over the family vacation she could have consulted a cousin who had done the same course at the same college, her parents and aunt and uncle could have all pitched in with ideas while we were relaxed and had time. But Maya was adamant that she wasn't going to start the essay until mid-September. She had her own timetable in her head and she wouldn't be swayed. I was frustrated. Why wouldn't she just do what we all suggested? But, sure enough, when mid-September came Maya started the essay and got it finished exactly on schedule. She knew what she wanted and four adults badgering her to do something different hadn't changed her mind." Katty now sees it as a strength that Maya didn't need to please the adults around her, even when they were annoyed with her. She had developed the confidence to listen to her own opinion.

The marble halls of the United States Senate sound different these days. Increasingly, the polished floors echo with the sharp tap of high heels. The arrival of a record number of women senators, twenty at the moment, means this bastion of shuffling, gray-haired men is slowly being feminized. In office 478, at the end of a long corridor in the Russell Building, we met one of the newer members, Kirsten Gillibrand, the junior senator from New York.

Everything about Senator Gillibrand smacks of confidence. She is impeccably groomed with a sleek, serious, Manhattan polish: Even her blond hair seems trained both to look stylish and to stay put. She already has her name on high-profile bills and her face on high-profile TV shows. Those in the know say she has presidential potential. The forty-eight-year-old mother of two is a Democratic star. In the calm of her pale blue office, Gillibrand, who first ran for the House when she was thirty-eight, confessed that she wasn't always like this.

"I didn't have the confidence to run for office until I volunteered on other people's campaigns for about ten years," she said with a laugh. What held her back was what she describes as those self-doubt questions familiar to so many women: "Am I good enough? Am I tough enough? Am I strong enough? Am I smart enough? Am I qualified?"

It's hard to believe that this woman, who has engaged in legislative battle with the military brass and taken on the gun lobby, ever felt she lacked grit or smarts. But then Gillibrand goes on to list all the effort she put into building her confidence for a congressional run: the years of unpaid volunteering, the night and weekend classes, and the voice coaching. Soon it became clear that she had made a deliberate choice to face off against her self-doubt.

Jane Wurwand, Christine Lagarde, and Elaine Chao started their self-assurance lessons young, from parents who unwittingly placed confidence-inducing responsibility on their small shoulders. But it's never too late to find it for yourself. Like Patti Solis Doyle, Caroline Miller, and so many of the women we talked with, Gillibrand proves that. She used the same menu we've been detailing—she took risks, she was persistent, she worked hard, and even failed. And it worked. Whatever she hadn't inherited, or soaked up as a child, she created.

There were a host of revelations for us as we researched this book. We didn't expect to uncover a clear genetic link to confidence. We didn't think the confidence gap would be as big as it is, or that female brains might work a bit differently, physiologically. One thing we had felt fairly sure about—that confidence is something largely acquired in childhood—turned out to be wrong. It never crossed our minds that we could feel distinctly mean as we tried to get better at "nurturing" some of it into our kids.

Our biggest and perhaps most encouraging discovery has been that confidence is something we can, to a significant extent, control. We can all make a decision, at any point in our lives, to create more of it, as Senator Gillibrand did. The science on how new behavior and new thinking affects, and literally changes, our brain is remarkable. Laura-Ann Petitto described the bridges and byways we can construct around the immovable concrete highway of our genetic code, and those extra lanes our upbringing puts down.

Caroline Miller and other psychologists contend that the volitional contribution to a trait like confidence may be as high as 50 percent. That means we ourselves, as adults, can make a decision to be confident, do the work, and see a result.

This idea of confidence as a choice opens doors in every direction. The temptation to say "I'm not good enough; I can't do it," exists for everyone at some point, in some circumstance. We've all heard "My mother didn't praise me enough," or "No one in my family is very confident." But when we write off confidence as purely a twist of genetic or environmental fate, we're shutting off possibilities that could change our lives. We don't need to be stuck in that pattern of self-doubt. It's a matter of pushing yourself to action over inaction, even in a man's world.

But here is where the path turns rigorous. You don't get to "choose confidence" and then stop thinking about it as your life miraculously changes around you. It's certainly not as simple as clicking a box to add self-confidence to your list of attributes. There's no glossy or beguilingly easy confidence prescription. When we say confidence is a choice, we mean it's a choice we can make to *act*, or to *do*, or to *decide*. If you've read only this chapter, you know that confidence is work, hard and deliberative, though we have no doubt that it is doable. And if you ease up? If you choose not to fully exert yourself to expand your confidence? That always-elusive image of ourselves in the mirror, the mirror that so easily shows men whatever they want to see, may never come into focus.

Consider, one more time, the work of Zachary Estes, the psychologist who conducted those computerized spatial tests on men and women. His results could not be any more straightforward, or any more relevant, for this issue. What held the women back was not their actual ability to complete the tests. They were as able as the men were. What held them back was the choice they made not to try. When the questions were difficult and the women doubted themselves, they held back. The men didn't have those internal brakes; they just went ahead and answered the questions as best they could.

If you choose not to act, you have little chance of success. What's more, when you choose to act, you're able to succeed more frequently than you think. How often in life do we avoid doing something because we think we'll fail? Is failure really worse than doing nothing? And how often might we actually have triumphed if we had just decided to give it a try?

Look around you. It's not usually a genuine lack of competence that holds most of us back from making that choice; the hurdle is a skewed perception of our abilities. It's as though we're wielding Adam Kepecs's forecasting tool, but one that isn't calibrated, that isn't giving us accurate readings, and is, therefore, dangerous. When we give in to negative beliefs about what we can and can't do, we don't seize the challenges we could easily handle and learn from. We aren't doing the basic stuff that would soon make confidence-creation almost automatic. But we have the power to recalibrate our confidence compass, and with it, our perceptions and our appetite for risk.

So, yes, as Sheryl Sandberg argues, we do need to lean in. We need to act, instead of holding back. And that means, we now know, that we have to be ready to work in ways that will often challenge our most basic instincts.

6

FAILING FAST AND OTHER CONFIDENCE-BOOSTING HABITS

One of our good friends (a male, an Internet genius, and start-up wizard) threw two words our way when we asked what he thought women should do to build confidence: Fail fast.

We laughed. As if! This was early in our pilgrimage, so failing still seemed exactly the opposite of what women do easily and naturally. Failure, it went without saying, we abhorred. And to do it quickly? That would mean we hadn't made a full effort, or done our work perfectly. I think we may have actually shuddered as we listened.

He wasn't joking. Fail fast, as it happens, is a techie buzz phrase, and more important, a hot business strategy. It's based on the principle that it's better to throw together a bunch of prototypes, roll them out quickly, see which one sticks, and toss the rest. These days, the world won't wait for perfection, and spending the time endlessly refining your product is just too expensive. Failing fast allows for constant adjustment, testing, and then quick movement toward what will actually work. The beauty is that when you fail fast, or early, you have a lot less to lose. Usually you are failing small, rather than spectacularly. And you have a lot to gain from learning as you fail.

We've come to see the theory of failing fast as the ideal paradigm for building female confidence. First, it just sounds more appealing than typical failure. It's not that it's "healthy" to fail, in the dreary way that kale cookies are healthy. It's actually hip now, even potentially lucrative. And this bite-size failing seems manageable. We need to fail again and again, so that it becomes part of our DNA. If we get busy failing in little ways, we will stop ruminating on our possible shortcomings and imagining worst-case scenarios. We'll be taking action, instead of analyzing every possible nook and crevice of a potential plan. If we can embrace failure as forward progress, then we can spend time on the other critical confidence skill: mastery.

Our quick failures will let us be choosy about how we spend our time. No longer will we need to try to get everything right. A lot will land in the garbage heap. We would do well to remember that it's not the strongest species that survives in the long run—it's the one that is the most adaptable.

Shortly after we talked to our tech friend, Claire decided to stick her toe in the failure water.

"One thing I've long wanted to try," she says, "is to give a speech where I simply wing it. To stand up there with no notes, and just talk. You know—Oprah, Ellen DeGeneres, Bill Clinton style. My gut tells me I'd be a more effective speaker; I'd connect more to the crowd and harness more of my energy. The fail fast concept pushed me to try. I didn't want to fail big and fast, of course, so I left half of my speech unscripted. It was a shock to hit that blank page. And

I wasn't great, to be honest. I got through it, but with much more meandering than necessary. A lot of *ums*. I'm not sure I covered the right ground. But I learned. A totally blank page doesn't work for me. Next time I'm going to try an old CNN live shot trick—just have a list of a few key words to guide me."

Confidence, as we've said (at least fifty times by now, and there are a few more repetitions to come), is about action. It also takes repeated attempts, calculated risk taking, and changes to the way you think. Sadly, you cannot simply square your shoulders, straighten your skirt, and ease into confidence as your grandmother might have suggested. Some of those vintage maxims can help, but not in isolation. The latest research yields intriguing, counterintuitive, and holistic implications for how to follow the confidence trail. A lot of the advice we'd never heard of before, and frankly, before we tried it, we just didn't believe it could work. But it does. We've ferreted out what we found to be the most salient information, especially that with application to our everyday lives. Grab the strands that resonate with you.

Leave the Comfort Zone

If you only remember one thing from this book, let it be this: *When in doubt, act.*

Every piece of research we have studied, and every interview we have conducted, leads to the same conclusion: Nothing builds confidence like taking action, especially when the action involves risk and failure. Risk keeps you on life's edge. It keeps you growing, improving, and gaining confidence. By contrast, living in a zone where you're assured of the outcome can turn flat and dreary quickly. Action separates the timid from the bold.

It's okay to start simple. If your confidence gap is in meeting new people, begin small: Pass food at a party and introduce yourself along with the salsa and chips, or make eye contact and then conversation with a stranger at the dry cleaner. If you aren't confident going to parties alone, try this: Start with a small gathering where you know there will be people who are friends, say yes in advance so you can't back out, then move on to the work reception, dare yourself not to back out at the last minute, and when you get there quickly find a group of two or three people and introduce yourself and ask them questions about their lives. Focus on their answers, be very present in that conversation; it will take your mind off the fact that you are there alone. If you aren't confident asking for a promotion, practice making the case on a trusted friend; give her five ways you've helped the department. Small steps prepare you for taking more meaningful risks. It's called the exposure technique.

And for many women, risk can take less obvious forms. Just as often it means allowing ourselves to be imperfect, braving the displeasure of authority figures and loved ones, or learning to be more comfortable when you're at the center of attention. Once you master these, you can build up to bigger risks: challenge your colleague's opinion on a project and don't cave at the first counterattack; try out for a play; take on a job that seems out of reach.

Sometimes the most important actions and risks to take aren't physical at all—they have nothing to do with talking in a meeting or applying for a new job. The ability to make decisions big and small, in a timely fashion, and take responsibility for them, is a critical expression of confidence, and also leadership, according to all of our most confident women. We listened to Linda Hudson give a compelling speech on the subject of decision making. Even if you make the wrong decision, she says, decide. It's better than inaction.

What's the worst that can happen, in all of these scenarios, when you leave your comfort zone? That's right. We're back there again. You could fail.

Beth Wilkinson is a quick decision maker, a consummate risk taker, and one of the most confident people we know. As an assistant U.S. attorney, she helped to prosecute Timothy McVeigh, and her consistent ability to win high-stakes courtroom victories has made her one of the most sought-after litigators in the country. Sometimes, though, even she fails. Indeed, she confessed to us, one early Saturday morning at Starbucks, that she's a pro at failing fast in small ways, which is the fallout from making lots of decisions quickly. She shrugs. "I usually learn from it," she says, laughing. One early failure has become a touchstone for her. It was one of her first solo cases and, wanting to get her opening argument just right, Wilkinson wrote it down, and read it verbatim, instead of trying to memorize it. Later, she overheard a male colleague criticize her performance. She was crushed. Instead of dwelling on it for too long, she thought it over, and realized that he was right. "It was a turning point for me," she says. "It is far better not to say everything perfectly, and to just connect with the jury. It taught me a lot. And I've never read a closing statement again." Here was an almost perfect execution of failing fast with a growth mind-set. Do it, learn, and move on.

Missteps really do provide accelerated opportunity for growth, as well as a chance to tap into that other internal resource we mentioned: self-compassion. As the research shows, practicing self-compassion provides a sturdy emotional safety net, one much stronger than our traditional concept of self-esteem. Self-compassion, you'll recall, centers on the acceptance of our weaknesses. Instead of saying, "I am not a failure," it's more useful to say, "Yes, sometimes I do fail, we all fail, and that's okay." It's extending the same kindness and tolerance—the very same qualities we find so much easier to afford our friends—to ourselves, while coming to terms with our own imperfections.

Taking a big risk, and surviving, can be life-changing. "My most confident moment coincided with my least confident," Jane Wurwand told us, because it meant finding the confidence to get out of a marriage that was undermining her confidence. She was young, living in South Africa far away from her family, in a society and era that frowned on divorce. She was worried about what would happen to her, socially and financially, if she abandoned the stability of married life, but the relationship was making her profoundly unhappy. Decades later, she still remembers the day she summoned up the confidence and courage to leave. And she remembers, too, the cheap plastic shopping bags her husband shoved her clothes in, before he threw them out of the apartment window. "That's all I left with, those two plastic bags, and I remember driving to my friend's house and thinking, 'I will never let myself be this vulnerable again.' It was the hardest thing I ever did. But it was also an incredible confidence builder. I thought, 'My God, if I survived that, I can survive anything.'"

Don't Ruminate—Rewire

Simply put, a woman's brain is not her friend when it comes to confidence. We think too much and we think about the wrong things. Thinking harder and harder and harder won't solve our issues, though, it won't make us more confident, and it most certainly freezes decision making, not to mention action. Remember, the female brain works differently from the male brain; we really do have more going on, we are more keenly aware of everything happening around us, and that all becomes part of our cognitive stew. Ruminating drains the confidence from us. Those negative thoughts, and nightmare scenarios masquerading as problem solving, spin on an endless loop. We render ourselves unable to be in the moment or to trust our instincts because we are captive to those distracting, destructive thoughts, which gradually squeeze all the spontaneity out of life and work. We have got to stop ruminating.

It's not easy. Even a neuroscientist sometimes can't help herself. Laura-Ann Petitto is a leader in her field. She's made numerous important discoveries about the origins of language, and runs a prestigious laboratory, supported by Gallaudet University and the NIH, which studies the brain and language development. She's created a new discipline known as Educational Neuroscience, won more than twenty international awards, and made an Oscar-nominated documentary about her pioneering work with a chimpanzee called Nim. When we met her at her lab, she was a welcoming and dynamic blur of energy and curiosity wrapped in a sleek orange and purple dress. Utterly confident, we imagined. Petitto told us though that while she knows she is thoroughly competent, she still dwells on her weakness—her fear of public speaking, for example. For years, she had a debilitating habit. She would sit on the bus on the way home from her lab creating a long list of her perceived failings. It was her mental default mode. "I could have done that better," she would say to herself. "That wasn't as good as it could have been. I shouldn't have been so nervous speaking in public."

Recently, she vowed to make a change. To break this negative pattern, Petitto decided to react to it by reminding herself of three things she'd done well. Now, when the negative ruminations start, she consciously goes through her list of achievements and successes: "That was a good paper I finished," the interior monologue might now go. "I got that lab report done quicker than I expected. I had a good conversation with my new grad student."

Such thought exercises rewire the brain and break the negative feedback loop. The effect may not be immediate, but in some cases, you can produce a regular change in thoughts and then in actions within weeks. You must start by becoming a keen observer of the relationship between your thoughts, your emotions, and your behaviors, and how one can affect the others. It is basic cognitive behavioral therapy. Here's one do-it-yourself exercise to help you become more aware of the link between thoughts and actions:

Think about a terrible scenario you imagine happening at work. Dwell on it. Perhaps you're giving a presentation, and you see coworkers rolling their eyes. Notice how you start to feel. Anxious. Stressed. Angry. Not great, right? Now, do the opposite. Envision something terrific happening at the office. An unexpected bonus. Nailing that presentation. Notice the feelings that wash over you now.

What we think directly affects how we feel. Even when nothing actually happened. Our mind did the work.

Kill NATs

NATs man the front lines in the assault on confidence, and they are every bit as annoying and insidious as their phonetic twin. We're talking about *negative automatic thoughts*. Unfortunately, they buzz around more frequently than positive thoughts, and can multiply at lightning speed. Do any of these comments sound familiar?

"That dress was too expensive. Why did I waste my money?"

"I'll bet when I go to work Sophie will be in earlier than I am again."

"I'll never finish this project; I knew I wasn't up to it."

"If I don't finish this project tonight, then I'll look bad to my boss, and I won't be promoted."

Unfortunately, you can't simply wipe out these NATs with a can of spray, but you can challenge them and wrestle them down with logic and alternatives. The first step is recognizing them. It may sound tedious, but keep a journal and write them down. There's no substitute for it. Just do it for a few days. Keep one by your bedside and jot down some of what's circulating in your brain each evening. Here's a sample from our notebooks:

Katty, during one evening:

I really need to ask for a pay raise, but will that make them mad? I don't want to seem too arrogant.

Why did the BBC call during dinner? With the book deadline I bet they think I'm not working enough. I should have called back.

I shouldn't have spent so much money getting the back of our house painted. No one can really see it.

Maya seems stressed by college applications, but if I hassle her she'll think I don't trust her.

Should I eat more, or less?

I'm going to be gone so much over the next three weeks. Is that too much strain on the house? I might need another sitter.

Claire, in the predawn hours:

Why do they let the planes circle above our house at 5:30 in the morning before landing? I really should call the community association and complain.

School—did the kids get their homework in their bags? I don't think I saw Hugo put his in.

How is there not enough time to get things done? I'm only working part-time right now for ABC—I should have more than enough time to handle the kids, and a book. I'm just not efficient. What is wrong with me?

I should really start getting up at 4:30 to write.

Are my arms flabby again?

I wonder when my husband is going to finally leave his high-stress job. That would sure help. He's just going to have to take charge of our vacation plane tickets. I won't have time.

I loved that picture of Della playing soccer. She's so strong. But is she practicing enough?

I think we are out of Rice Krispies.

(We laughed as we compared our notes, half-mortified. Our musings seemed just too embarrassing to share. Ultimately, we decided we'd be doing a public service.)

The best way to kill a NAT isn't to beat yourself up for having it. That simply leads to more anxiety. The most effective and surprisingly easy fix is to look for an alternative point of view. Just one different interpretation, perhaps a positive, or even neutral, reframing, can open the door for confidence. So, since we are offering ourselves as guinea pigs, here are a few of our own attempts.

> "I'm just not efficient, what's wrong with me," becomes "Maybe I am doing a good job balancing so much, actually."
>
> "Why are the bosses calling now?" becomes "Maybe they want me on television more, and that's a good thing."
>
> "Did I spend too much on getting the house painted?" becomes "We had water damage and this should help, that's why I did it, after all."

The second thought doesn't even have to prove the first wrong. It's the mental exercise of taking the time to create another explanation that can lessen the potency of the first thought. Eventually, reframing becomes a habit. And if you're struggling to come up with positive alternatives by yourself, imagine what you would tell a friend who confessed to having that same negative thought. This is putting self-compassion into action. You'll be surprised how quickly you can trim those debilitating feelings down to size. It's easy to do for others, yet we let them roam freely in our own brains.

Richard Petty's research suggests that getting physical with your thoughts can also help kill NATs. He and his collaborators asked a group of students to write down bad thoughts about themselves; they then divided the students into three groups. One group was instructed to put what they'd written in their pockets and carry the notes around with them. The other group was told to tear up their notes and throw them in the trash, symbolically exorcising them. The third group was instructed to leave the pieces of paper on the table.

"It turns out this symbolic interaction with your thoughts affected how correct you thought those thoughts were," says Petty. The people who carried the paper around with them grew more concerned about the negative thoughts, as if they had some value. The people who threw the notes away started questioning the validity of their negative thoughts, and soon their thoughts didn't bother them. And the people who left them on the table were somewhere in between.

These strategies help build firewalls that keep toxic thoughts in check. If you are rejected for something, it doesn't mean you'll never be successful. If you get negative feedback on a piece of work or task you did, it doesn't mean you can't improve the next time. If you are nervous about a big interview, don't dwell on the possible outcome and leap to the conclusion you may never work again in the industry if you don't get the job. Attack the concepts with your new tools. Counter them with facts—and then toss those negative thoughts aside—even if it sometimes means you have to throw that notebook we told you to keep in the trash.

Our attention is a powerful force, and it actually is not hard to use it to our advantage, it turns out. Sarah Shomstein, a neuroscientist at George Washington University, told us researchers are coming to see that the simple act of thinking, of *focusing*, on almost anything—the new car you want, exercising, your project—means you are likely to take action in that direction. We need to make our thoughts an ally.

From Me to We

You might think that focusing more on yourself would be the natural stepping-stone to confidence. Don't we need to make it all about ourselves to feel good, to succeed? Actually, the opposite is true, especially for women. For most of us, thinking about our feelings and abilities, judging ourselves, and making ourselves the stars of our own melodrama, tend to inhibit and paralyze us. Imagine this, and you'll see what we mean. How might you behave in an emergency, pressed to save a child? There'd be no time to be nervous or to second-guess your actions. You wouldn't stop to ask whether you were qualified or whether you should perhaps take another course in CPR before you jump out into the street. Your attention would be placed entirely on averting a crisis, and you would excel without a moment of doubt.

Now, apply that same thinking to your own challenges. If you have a big event approaching, for example, at first it may seem natural, and even helpful, to think and think and think about it, to examine the situation from every potential angle and to prepare for every possible scenario: what it might mean for you in the long run, how you will look, what you should say, what you should wear depending on the weather, how you will handle every possible contingency that could arise. That's not the way to do your best work. Instead, do the prep, and then turn your attention to how much it will help the team or the company. That will liberate you to be bold and assertive and to redirect the spotlight.

OSU psychologist Jenny Crocker has found that women thrive on *we*. When young female college graduates, whose confidence is wobbly, stop thinking about how they can prove themselves and move instead toward doing things for colleagues or the enterprise,

Crocker found they get a surprising boost of confidence. She's used that research to develop a great tip for nervous public speakers: Reframe your remarks in your head. Tell yourself you are speaking on behalf of the team, or the organization, or for the benefit of others, rather than for yourself. Change some of your language if you need to. It's a simple, practical way of moving that spotlight off yourself and onto others to give you confidence.

Senator Kirsten Gillibrand uses similar principles to persuade women to run for Congress. She reminds them that this is much more about helping those who need protecting than it is about them. "As soon as a candidate realizes it's not about self-aggrandizement, particularly a female candidate, they become stronger and they become more purpose-driven," she says.

It's Not Personal

It's a whole lot easier to move from *me* to *we* when you realize other people aren't actually thinking about you all the time. Out of some misguided narcissism, it's all too easy to think that whatever you have done—whether it's a triumph or a failure—is the focus of everyone else's attention. It isn't. Most people are too busy getting on with their own lives to worry about what you're up to. Imagining that you're the center of everyone else's universe is silly, and it kills confidence. When you didn't get elected class president, or when you made a mistake in a client meeting, don't think anyone is gossiping behind your back for weeks on end. They aren't—they moved on a long time ago.

When you do face a problem at work, remind yourself that it's about the work, not about you. If your boss tells you a project you've been working on needs to be improved, resist the temptation to see it as a personal attack. When your colleague asks how your weekend was, without a smile, realize it isn't a jab implying you should have been in the office. Really, it's the ultimate in egotism when we think like this:

> "I'm sure she's angry because I didn't add that point she suggested."

> "I know he must think I'm an idiot because I still haven't set up that meeting."

Put that alternative thinking to work:

> "On the other hand, she's/he's got four meetings today. I doubt it's an issue."

In some cases, comments and critiques are meant to be personal. And in some professions, the judgment is constant. Performers, for example, live with a different level of scrutiny.

"In theater you are criticized from head to toe, from your eyebrows to your earlobes, from costumes to makeup—you can go crazy," says Chrissellene Petropoulos, an opera singer and voice coach. "You're never being told wonderful things; you're always told terrible things, and I used to choke and fall apart because I was personalizing everything. The conductor would come up and say, 'You don't know how to sing,' and 'You look like an elephant,' and I was like . . . ahhhhhhhh!"

Petropoulos realized that the way she was experiencing the feedback was destroying her performance skills. She started to study the impact that stress has on vocal cords and was stunned. She learned to interpret criticism as something directed at her skills, not her value as a person.

Today, she's in huge demand as a teacher, and her lessons focus on confidence as much as on voice skills. She rehearses long lists of rote responses with her students, many of them children, so that they are ready to handle and process critiques.

Critique:	Your hair looks horrible today.
Reply:	Thank you for saying that. Or, thank you for noticing.
Critique:	You're singing through your nose again. Stop it. It sounds terrible.
Reply:	Thank you for telling me that. I'll try to do better.
Critique:	That outfit isn't going to work.
Reply:	What would you like me to wear, or how do you think I should change it?

Petropoulos says her young students may giggle about the way those comebacks sound, but in the long run, they become more mindful about how they receive and internally process negative information. Claire has found thinking *thank you* before making any retort when she feels criticized has helped her fight personalizing tendencies. You can come up with your own response list appropriate for your needs: *Thanks for the feedback. I appreciate that thought.*

If you just can't break your personalizing habit, a big dose of reality always helps. Remember that a lot of other people face exactly the same hurdles that you do, and that for all women, many forces beyond our control will affect our careers. "It was really enlightening for me in terms of my confidence when I finally realized, 'Okay, here's a lens to help me understand some of these obstacles that I've experienced,'" says Christy Glass. "It was a language for saying, 'It's not that I can't do this job because I don't have the skill. It's that the resources I need to do this job have been denied to me. It's not always that I'm just not aggressive enough, or that it's an individual failure.'"

She says just being aware that there are indeed workplace biases is a powerful antidote to self-doubt, especially for young women who might not remember *Ms.* magazine. So, the next time you walk in to give a presentation to an executive board and see fourteen men and two women around the table, as Katty recently did, realize that a slump in your confidence is to be expected, based on larger forces. Even that recognition can help you move on and not beat yourself up for feeling a bit nervous. It doesn't mean you need to dwell on the unfairness, and you certainly shouldn't give up, or complain incessantly, but understanding context and institutional dynamics can help you keep disappointments and challenges in perspective.

When We Should *Star in Our Own Production*

Often, women just seem to have the spotlight thing *backward.* We want to shine a bright light on our faults, insecurities, and the outlandish reasons we will surely fail, but when it comes to taking credit or enjoying our triumphs, we step into the shadows, looking askance at our accomplishments as though we've never seen them before. There are plenty of times when the focus *should* be on us, when we need to move from we back to I. You do need to develop a sense of your own, well-deserved value to the enterprise and, yes, sometimes you even need to toot your own horn. It can help your case at the office, but also just the simple act of doing it, of hearing ourselves recognize our accomplishments, bolsters confidence.

For most of us, being self-deprecating seems far more appealing than boasting, but that can backfire on multiple levels. Even if we're simply trying to downplay achievements in front of others, we are essentially telling ourselves a damaging story—that we don't really deserve our accomplishments. That affects not only how we see ourselves, but also how others see us. Remember, our bosses want winners working for them. They like to hear about what we've done well. Moreover, if we devalue, to ourselves, what we've already achieved, it makes it less likely that we'll attempt to clear future hurdles.

We have to find ways to take in compliments and own our accomplishments rather than relying on dismissals and assertions of luck and self-deprecation. Keep it simple if you must. When praised, reply, "Thank you. I appreciate that." Use it. It's surprising how odd, and how powerful, saying those five words will feel.

Both of us were mulling over all of the possible examples of our own self-deprecation, when our editor, Hollis Heimbouch, was quick to note a particularly apt one. Even in this very manuscript, we were unable to keep from poking fun at ourselves in various asides: Our scientific chops, our business sense, and our organizational ability were all targets (good-natured, we'd thought). It's a habit so ingrained we didn't even notice it. Fortunately Hollis did, and quickly pointed out that a few years of research and writing, not to mention professional speaking on the subject, meant those self-directed barbs weren't even believable.

Repeat, Repeat, Repeat

Michaela Bilotta grew her confidence along with her powerful biceps one painful pull-up at a time. You don't get to graduate from the U.S. Naval Academy without being chiseled and fit. That was never daunting for Bilotta. She thrives on ferocious workouts, except for the pull-ups. Bilotta hates them, and she measures her own standing against her contemporaries by how many pull-ups they can do in a row: "So-and-so's a twenty," Bilotta says of a classmate. "She's awesome."

Mastering pull-ups has meant hours of effort and persistence over the past five years. But it's paid off. Now Bilotta's a fourteen or a fifteen, and she feels proud of that, and confident about her ability. "I have to work at it, work at it, and work at it to get to my fourteen. But if I wasn't willing to do that then I should not have chosen this service."

It's the same with confidence. You won't get it if you don't work at it, because all of our self-generated confidence comes from work, and mastery in particular. (We'll caution again, that by mastery, we do not mean you should unleash perfectionism. Think "good enough" as you conquer new frontiers.)

No one better illustrates the mastery-confidence continuum than Crystal Langhorne of the Washington Mystics. "My first year I didn't play at all," she says. "And I didn't play well at all. At the end of the season, I was wondering if basketball was even for me anymore." Langhorne grew quiet for a minute as she remembered one of her toughest professional experiences. Instead of quitting, she told us, she came up with a different plan. More practice. Not just a bit more, but hours of shooting, every day after practice, while she was playing ball in Lithuania during the off-season. She knew she had to completely remake her shooting style. And she did.

The impact was not just noticeable, but remarkable. Crystal was named the most improved player in the league when she returned, and she's been voted an all-star player every year since. She literally changed her game by staying in endless motion, shooting a ball at a basket.

Practice. A willingness to learn. Those are now her go-to confidence boosters in the games. "When you work on things you think: 'I know I can do this. I worked on it. I did it in practice.' It gives me confidence." Another reminder that the people who succeed aren't always naturals. They are doers.

Speak Up (Without Upspeak)

The idea of talking to a group of strangers hovers like a dark cloud over most everyone's confidence, and public speaking is an iconic challenge to female self-assurance. Running Start, the group that advises young women considering a run for political office, has identified the fear of speaking publicly as the number one thing that stops women from getting involved in electoral politics. But because most of us have to do it at some time or another, it's worth tackling this one. Whether you are in a book club or a boardroom or at a birthday party, at some stage you will need to make your thoughts heard, and that means being able to speak in public with assurance. Like so many other things, it is a learned skill. If you know you can master it, even in an elemental fashion, the confidence you gain is profound.

Let's take the annual conference. You're there with thousands of colleagues from your industry, and you are in a packed auditorium to listen to the keynote speaker. At the end of the forty-five-minute speech, there's definitely something you'd like to ask. But when the speaker inquires whether there are any questions, not a single female hand goes up. The men, confidently, dominate the Q&A session. While the women sit there, mute, thinking—what? That they will seem stupid, ill-informed? That they might stumble? That everyone will stare at them?

This is not made up. Remember the studies we mentioned in chapter 1? When men are in the majority, women speak 75 percent less. We both give speeches, and we see it time and again. If it's an audience largely made up of women, it's different; women usually have no qualms asking questions in front of other women. But when we give talks to male-dominated audiences, or mixed crowds, the women always seem to struggle to make themselves heard. Katty says she was amazed to learn that even her uberconfident sister Gigi is one of them. Gigi is a world-renowned veterinarian, one of the few dealing in donkeys and mules. In fact, she's often asked to give speeches about her work with equines and she really enjoys it. She has no problem taking a stage before an audience of several hundred peers. Recently, though, she noticed something really odd. When it comes to attending other people's speeches and asking questions, no longer the expert, in her mind, she gets really nervous and has to force herself to raise her hand. It makes no sense.

We were both bowled over by the confidence of a young woman we met socially at a dinner recently. Somehow, what had been an easygoing conversation turned into a combative argument about, of all obscure things, women and their place in religion. A man at the end of the table kept insisting he was right. One by one, the other guests gave up arguing with him, but the twenty-eight-year-old refused to be intimidated. He was almost twice her age, but she wasn't remotely cowed. So often, women in social settings, even more than in professional arenas, back down at the first sign of a conversational challenge, but she had no problem arguing her position, and then sticking to her guns. She wasn't rude, indeed she was charming, but she didn't give up. It was impressive.

The ability to advocate for ourselves in smaller office settings or around dinner tables prepares us for those critical moments when we need to speak to a huge crowd, or maybe just to an important audience of one, asking for a better deal.

Wherever it is, projecting yourself effectively is a constant test of confidence. Often you have to steel yourself, overcome your natural self-consciousness, and command your vocal cords to follow your will. But think it through: When you do, what actually happens? Worst-case scenario, you blush and mix up your words; maybe dark circles appear under your arms. But the ground does not open to swallow you whole. The sky does not fall on your head. No, you're still there, intact, and alive. There are loads of wonderful books on speaking, so we won't hijack that specific advice. But we'll share with you a few points we came across in our confidence excavation that are new and useful.

First, use your own style. You do not need to emulate Nikita Khrushchev, shoe in hand, pounding on a desk. Peggy McIntosh, the Wellesley professor, writes persuasively about something called our "home selves," a state where women really do feel in command. When we can bring that level of comfort, and that style, to our professional lives, even though it may not feel pinstriped, she believes we project more authority.

Second, you will feel power speaking on behalf of others, as OSU psychologist Jenny Crocker found; so use that as a tool for crafting your remarks. A focus on the lofty goals, or the accomplishments of the team, will imbue your performance with a natural sense of mission.

Finally, banish upspeak. Christopher Peterson was greatly loved by his students in Ann Arbor, where he taught psychology at the University of Michigan for years and was honored with the Golden Apple Award for outstanding teaching. He was also one of the founding fathers of positive psychology. He died suddenly, at the end of 2012, but we were fortunate enough to have absorbed some of his wisdom in an earlier interview. He had a pet peeve. Peterson hated the way so many of his female graduate students talked. Time and again he'd be in lectures, listening to very bright young women answer questions, and they'd be using what he called *upspeak*. It's that habit we know you'll recognize (and perhaps even suffer from yourself): raising the tone of your voice at the end of a sentence in a way that suggests what you are really doing is asking a question, not making a declaration. Read this out loud—"We went to the movies, and then we got ice cream." And now this one—"We went to the movies? And then we got ice cream?" Or worse, try this one—"I think we should go with the online marketing strategy?" Awful, yes? It may come as no surprise that linguists report upspeak is most common among women in Southern California. But it's now Valley Girl gone mainstream. Researchers say the questioning style serves a clear purpose for women: It is a psychological safety net; it discourages interruptions and encourages reassurance. So, when we're unsure of ourselves—not because we don't have the knowledge, but because we are nervous about sticking our necks out—we unconsciously make our comment sound like a question in order to deflect criticism.

Upspeak made Peterson cringe because he heard it as a conversational hedge that revealed a lack of confidence on the part of his female students. He described it as a way of saying, "Don't challenge me because I'm really not saying anything; I'm just asking."

He told us most of those same grad students had great potential, but that the upspeak was so distracting it was getting in the way of their progress. When they always seem to be hedging, it detracts from the validity of their argument.

Peterson found no evidence of this pattern in men. If anything, his male students erred on the side of overconfidence. They could be abrupt and direct, and prone to wagging their finger in the air. He wondered if he should just let the issue slide, but he saw it was holding back his female students and that it could be easily fixed. Upspeak, after all, is not hardwired in a woman's DNA. So he was always, good-naturedly, nagging the upspeakers.

Here's the wonderful advice Peterson gave shortly before he died: "Say it with confidence, because if you don't sound confident, why will anybody believe what you say?" After that interview both of us were appalled to hear the occasional lilt in our own sentences—we weren't even aware we were doing it—and in those of our daughters. It's something we don't hear in our boys. "Say it like you mean it" has become our mantra for our girls and for ourselves.

Micro-Confidence—Dos and a Don't

The big confidence habits offer a broad prescription for getting more self-assurance. Make them yours, and you can even rewire your brain to become more reliably confident for the long term. But sometimes quick fixes can help. We have uncovered some small-scale, granular wisdom and quirky tips worth sharing, and we've unearthed one old confidence chestnut you should try to avoid.

- **Meditate.** A calm brain is the ultimate confidence tool, and meditation is so common and valuable that it's being taught in some of the military's basic training courses. Remember what we wrote about how much healthier a brain looks on meditation? It is literally rewired. Your fear center, the amygdala, shrinks. You have an increased ability to control your emotions and to be clear, and calm, about your goals. Claire tries to do it regularly, though she often fails (not the sort of failure we're after). When she does do it successfully though: "I have such a calm power over my acrobatic thoughts it's amazing."
- **Be grateful.** New research shows that gratitude is one of the keys to happiness and an optimistic mind-set. Find it in the tiny things: As someone lets you merge into traffic, notice, and be grateful, instead of just zipping ahead looking for your next maneuver. And again—just say thank you. Believe and be grateful for the kind words said about you. It will transform your mood, and simply saying, "Thank you, I appreciate that" will also make the other person feel good.
- **Think small.** Battle feelings of being overwhelmed by breaking it down. Teasing out the individual parts of a challenge, and accomplishing even one-tenth of it, can give you a confidence boost. "I'm a very logical thinker. My degree is in systems engineering, which is all about taking complex problems and breaking them down to their component parts," says Linda Hudson. Simplifying all that lies before her is what helps Hudson confidently solve problems. "When something's daunting, even in my personal life, I say, 'Okay, let's break this down into pieces I can manage and take it one step at a time.'"

- **Sleep, move, share—in any order.** Yes, we sound like your mother, but it's true. A lack of sleep and exercise produces an extremely anxious brain. (We've tested and retested the theory, and there's no getting around it.) And being close to our friends boosts our oxytocin levels. So indulge in guilt-free girlfriend time.
- **Practice power positions.** Sitting up straight will give you a short-term confidence boost, according to a recent study conducted by Richard Petty and his colleagues. Try it now. Abs in. Chin up. Astonishingly simple, woefully infrequent. Try nodding your head. You feel more confident as you talk when you do it—and you're sending a subconscious signal that makes others agree with you. And, yes, always sit at the table. Otherwise, you're handing power away by not sitting with those who have it.
- **Fake it till you make it.** Okay, here's the one to avoid. Attempt this bit of pop psychology at your peril. Originally an observation made by Aristotle, "Men acquire a particular quality by constantly acting a certain way," the modern version has become tainted by its suggested swagger, and if performed in that fashion, can easily go wrong. The very notion of straying far from our real selves is at odds with the central premise of this book. Confidence isn't about pretending, or putting on an act; it springs from genuine accomplishment and work. Jenny Crocker told us, in fact, that not only does faking it not work as a confidence booster, but it almost certainly makes us feel less secure, because knowingly masquerading as something we're not makes us anxious. Moreover, as good as we might be at faking it, we'll certainly project those

subtle false signals described by Cameron Anderson, and that won't help us much, either.

The appeal of faking it, if only for a while, is that it offers a crutch—a way to begin. Here's a better way to reframe the premise for a quick confidence jump-start: Don't pretend to be anything or anyone—simply take action. Do one small brave thing, and then the next one will be easier, and soon confidence will flow. We know—fake it till you make it sounds catchier—but this actually works.

- **Reach for the bottle.** When all else fails, you can always use the oxytocin spray. We tried it. Our husbands seemed sweet, and our work and children felt manageable. Modern liquid confidence! On the other hand, we'd also exercised and were sitting up straight so they were, we admit, fuzzy results from an unscientific study.

7

NOW, PASS IT ON

When Jim Stigler was a graduate student in psychology, he flew to Japan to study different teaching methods. One day, he found himself in the back row of a math class full of ten-year-olds. The teacher was trying to get the children to draw three-dimensional cubes, and one child was really struggling, producing shapes that looked deformed. The teacher called that child up to the front of the room and asked him to draw his design on the board. That surprised Stigler. In American classrooms you wouldn't single out the child who *couldn't* do something. That would be seen as humiliating the poor kid even further.

The young Japanese boy started drawing in front of everyone, but he still couldn't get it right. Every few minutes the teacher would turn to the class and ask what they thought of his efforts, and his classmates would shake their heads, "No, it's still not correct." As the exercise continued, Stigler noticed that he himself was getting anxious, and started sweating. "I was really empathizing with this kid," he says. "I thought, 'This kid is going to break into tears!'" But the boy didn't break down. He just kept on going, calmly, with determination. And, eventually, he got it right. The whole class broke into applause as the boy sat down with a huge smile, proud of his achievement.

Stigler, now a professor of psychology at UCLA, has come to the conclusion that the profound difference in the way the West and the East view learning has a big impact on confidence. It all has to do with effort. In America, Stigler says, "We see struggle as an indicator that you're just not very smart. People who are smart don't struggle; they just naturally get it. In Asian cultures, they tend to see struggle more as an opportunity."

There is a lesson here for all parents, and everyone who is in a position to guide young minds. In writing this book, we have come to believe that confidence is one of the most important qualities we can instill in our kids. But we are not talking about recycling that well-worn, clichéd technique—in which parents tell their children, try to convince them even, that they can be anything they choose to be. It sounds good. But kids recognize that as an empty assertion. They are creatures hungry for tangible proof.

Confidence gives them something else entirely: a faith in their ability to make things happen, to risk failure, and to all the while maintain a sense of inner calm and equilibrium. Confidence puts meaningful tools in their hands, instead of unproven promises in their heads. It won't guarantee success, but, more meaningfully, it lifts self-imposed limits. That's what we want for Felix, Maya, Jude, Poppy, Hugo, and Della. And it's a possibility, an advantage, all parents, whatever their faith or culture or economic status, can create for their kids.

Praise Progress, Not Perfection

In Japan, Stigler found that struggling with and then overcoming hurdles becomes a chance to show that you've got what it takes to succeed. In Japanese classrooms, teachers routinely assign children tasks that are slightly harder than what they've already been taught, in order to let them struggle with something that is just out of reach. Then, once they've mastered it, the teacher helps them see how they were able to accomplish something they thought they couldn't, through work and fierce effort.

Teaching a child to accept and even embrace struggle, rather than shy away from it, is a crucial step along the path toward instilling confidence. You are showing the child that it's possible to make progress without being perfect.

We've mentioned the self-esteem movement backlash. Psychologists worry that we still haven't learned the lessons it should have taught us, that we still aren't letting our children struggle enough. They fear we are bringing up a generation of narcissists, young people who've been told they can do no wrong and so see no need to improve. Jean Twenge, a professor of psychology at San Diego State University, sounds the alarm about millennials who have been raised by parents eager to reward their child's every move with that infuriating ubiquitous phrase *"good job!"* These children, says Twenge, are attention seeking, have a disproportionate focus on appearance and status, and may even have difficulty forming strong relationships.

You may recognize a few of them. These are the kids who played in soccer leagues where everyone was a winner and no one was allowed to lose. They were the generation who got a trophy just for showing up to the basketball game—the kids whose doting, hardworking, guilt-ridden, baby boomer parents believed that telling their children they were perfect was the best antidote to the harsher discipline of their own parents. Plus, relaxing the rules and lowering the expectations bar for their kids seemed to make up for those long days in the office.

But when we tell our kids they're already perfect, we are encouraging them to avoid things they find hard. And how do you cope with failure as an adult when you've never been allowed to lose in Little League? The cycle of losing, coping with loss, and then picking yourself up to try again is an essential component of mastery, not to mention confidence.

That doesn't mean all praise is bad. Psychologist Nansook Park says parents should simply make the praise specific to a task and as precise as possible, especially with younger children. For example, imagine asking your four-year-old to help set the table. When he follows your instructions to put out the spoons, Park suggests that saying, "Oh, you are the best son in the world" is too generic. "You have to help them to own what they did," says Park. "So, say something like, 'Oh, I like the way you have put the spoons on the table.'" And if he gets the spoons mixed up with the forks and knives, who really cares. The important thing is that he tried.

So, let your children make a mess of the cutlery, fall off bikes, crash from monkey bars. And for your part, stop getting overwrought about it. How *you* react can help build a spirit of independence and an aptitude for risk in your child.

Katty has always prided herself on pushing her children to stand on their own two feet; she sees it as one of the biggest differences between her British friends and her American friends. Americans are more protective and involved in the minutiae of their kids' lives. Brits are more laissez-faire. They haven't quite shaken the Victorian philosophy of children being "seen but not heard." But when she looks at how little she has actually let her children fail, she's shocked by how hard it has been for her to apply this advice. "They sometimes came to me in a panic fifteen minutes before school pickup, in tears, saying they'd forgotten a key piece of homework. Part of me is always tempted to say, 'Too bad, you should have remembered yesterday evening when you were busy watching TV.' But, inevitably, I'm racked with unhappiness at their tears, and whip out a pencil and 'help' them get it done in time."

We can't bear our children's suffering so we fix their problems, in school, in athletics, in their friendships. But over the long term they become too reliant on us and accustomed to bad things simply being swept out of their way.

How to Fry an Egg

Whether the cause is overprivilege or overprotection, many of us haven't taught our kids to cope very well with the basic challenges of life.

Dermalogica founder Jane Wurwand is refreshingly frank about how she's indulged her two children, admitting that she fears she has done them a disservice. But it's never too late to have an impact, and she's come up with a simple remedy: Start small.

"It doesn't have to be learning ballet or Chinese, it can be much smaller," she says. "My kids went to smart private schools, but they didn't learn to polish their own shoes. I missed out in not teaching them they can do basic things by themselves. We should draw up a list of twenty small things our kids need to be able to do to cope with life."

Here are a few things on Jane's list. You'll see it's not hard to come up with your own.

Call a friend instead of texting them

Do your own laundry

Take the bus

Fry an egg

Sew a hem

Change a button

Confront a friend rather than posting comments on Facebook

Make the experience fun. Make it a game if necessary (though don't reward it with a prize). However you approach it—one egg, one button, one bus ride at a time—teach your kids that they can master the basic skills in life. And here's your real challenge: When they botch the test or burn the dinner or miss the bus, don't jump in to fix it or get angry. For all of us, mastering skills requires the ability to tolerate frustration, and if you respond too quickly with a helping hand or agitation, your child won't develop that tolerance. Take a deep breath and let them figure it out. Let them fail.

It's Not About You

In Ireland, the Department of Health and Children sought recently to quantify the impact parents have on their child's mental well-being. They described positive mental health as "being confident in who you are" and being able "to cope and deal with things"—both key attributes of confidence. They asked young people across the country what they felt hurt their mental health. On the hurt list: people judging how they look, the pressure of school and exams, and family dynamics. One of the family factors these young people found most damaging? "Being expected to live up to parents' expectations since parents sometimes want you to live their dream." As we encourage our children to try new things and take risks, we must be careful that we're doing it for them, not us.

We see this all the time, the parents whose own status seems to depend on how well young Henry or Hannah is doing. Those are the parents who embarrass their kids by getting into screaming matches with the referee on the soccer field. Or, better intentioned but potentially just as damaging, these are the parents who spend long nights drilling their teenagers before a critical exam, telling themselves they are merely coaching, when the truth is that they can't face the prospect of their child not being a high achiever. And we've all read the horror stories of the parents who turn up at their child's first job interview.

What *is* about you is what your children learn from watching you, what they get from the example you set. When they see you struggle and prevail, or simply work hard, your children absorb it. Tanya Coke, our accomplished lawyer friend, thinks this is one reason many African-American women have a confidence habit to fall back upon, nurtured by their mothers.

"Black women of my generation grew up used to taking care of business," she says. "We had to. Most of us grew up in families with mothers who worked outside of the home. I can't think of a single black friend of mine whose mother didn't work. So our model was strong—we do what we need to do to support our family economically. We don't question our need to get out there and lead if necessary. But that doesn't mean it's not hard once we're there, of course."

Raising Confident Daughters

Many of the lessons we've gleaned about confidence building apply to all of our children, but some are specific to our daughters. When it comes to instilling confidence, raising girls to be more assertive and more independent takes conscious effort, and it goes hand in hand with encouraging them to be less good.

It begins innocently enough. What harried adult, parent, or teacher doesn't appreciate the child who is helpful, quiet, and generally well-behaved? Let's face it—these low-maintenance kids are just easier. It's not that any of us are knowingly pushing the idea that girls should be good; it's just that girls have an easier time pulling that off at an early age. As we discuss in chapter 4, the result is that young girls, consciously or not, quickly learn that kind of behavior is a fast track to praise. Soon, it's a reward cycle that's hard to break, and the result is that we subconsciously train our daughters not to speak up and demand to be heard, or demand almost anything. By the time our focus shifts to that, habits are hard to break.

We aren't suggesting that we should cultivate belligerence in our daughters, but this constant cycle of pressure and reward for good behavior doesn't help girls feel confident later in the rough-and-tumble world of the workplace. The impulse that lets many boys shrug off nagging parents, break curfews, and refuse to take showers is the same impulse in adulthood that inures them to the fear of annoying their bosses by asking for pay increases and promotions. They worry less about upsetting their superiors because, unlike their sisters, they haven't been trained to fall into line, and their brains aren't wired to be as sensitive to criticism.

When you are an overstretched parent—let's face it—having a daughter you can count on to be the good child can make your own life much easier. But if you want your daughter to have the confidence later in life to buck the system and advocate for herself, you need to encourage her to be a little bit bad.

It's a two-step process. First, don't overly criticize the bad behavior. When your precious girl does interrupt, shriek, throw a tantrum, or tear her new dress, check your instinct to reprimand her. And especially check your instinct to tell her she's acting out of character, as if somehow being the golden girl was what she's supposed to be. Phrases such as "Mary, I'm so disappointed, it's not like you to cause a fuss/not help/be naughty" need to go.

Second, don't overpraise the good behavior. This seems counterintuitive, almost wrong, but it's just the flip side of trying to get our girls out of the habit of feeling they always have to be ideal. Because if you constantly reward your daughter for helping out, keeping quiet, or being tidy, you're instilling a psychological addiction to goodness and to the praise that follows it.

Remember, you never know how powerful that independent behavior will turn out to be. Listen to former Washington, DC, schools chancellor Michelle Rhee, a woman who single-handedly tried to reform one of the worst school districts in the country. She battled the unions, pissed off parents, and never seemed to care. That gave her considerable power. "I don't give a crap if people like me or not, and apparently I never have," says Rhee, laughing. During the height of the DC public schools upheaval that Rhee oversaw, when she was getting hammered in the press day after day, her mother came to stay with her. One day her mom turned on the television and saw footage of people screaming at her daughter during a school board meeting. She turned that off and opened the *Washington Post,* only to find a two-page spread with more of the same vitriol from parents and teachers. When her daughter got home that night, her anxious mother found her in the kitchen making a peanut butter sandwich. "She comes into the kitchen and whispers, 'Are you okay?' I said, 'Yeah, I'm fine,'" Rhee recalls. "And Mom said, 'You know when you were young you never used to care what people thought about you.' She said, 'I always thought you were going to grow up to be antisocial, but now I see that it's serving you well.'"

We've also come to realize that confidence won't look the same in every child. Katty has two daughters, Maya and Poppy, who are living proof that self-assurance can express itself in very different ways.

Maya, a teenager, is amenable and helpful, but she's also really determined, even quite stubborn. She's definitely a leader who, in her own nonconfrontational way, won't be swayed by her peers or her parents. Whether it's dating or drugs, she's very sure about what she wants for herself and feels no need to go along with the group. Maya's confidence is quiet but solid. Poppy, her younger sister, is confident in a completely different way—much more in-your-face. "I'd never had a child who said *no* with such determination until Poppy was born," says Katty. "My other three were all fairly easygoing. But Poppy doesn't care what people think of her—not me, not her teachers, not her older brothers and sister. She isn't interested in pleasing anyone. If she's mad at you, she'll say it. If she doesn't like someone, she won't hide it. If you suggest a plan she doesn't like, she'll just say no. She has zero problems expressing all her emotions, all the time. It's sometimes exhausting and it's certainly demanding, but it's undoubtedly confident."

Discourage Pointless Perfection

Striving to grab the good-girl ring as a child sows the seeds of trying to be perfect as a woman. Girls internalize the lesson that they need to get everything right to reach the top of the class, which leads to perfectionism. But this ends up smothering true achievement. Perfection is the enemy of the good. It's also the enemy of confidence.

The danger is particularly acute for all those high-achieving girls. In her book *Supergirls Speak Out: Inside the Secret Crisis of Overachieving Girls*, Liz Funk describes how many girls now take the challenge of being extraordinary so far that they push themselves to the breaking point.

Overachieving girls might think they are hitting all their marks by working all hours, but they're actually not doing as well as they could if they just eased off a bit. When they move on to the workplace, these girls are the ones who will take on too many projects, because they believe they're the only ones capable of doing them well. They become so focused on getting the day-to-day tasks exactly right that they don't take time to lift their eyes and look at the big picture. Convinced of their path, they often become impossible to challenge, and eventually alienate their peers, all the while, failing to progress. That isn't confidence; it's a myopic and isolating self-righteousness.

Here are some ideas for discouraging perfectionism in your daughter:

- Praise her moderately, not excessively. Saying "Well done for working so hard on this" is much better than "You are the best student ever."
- Help your daughter feel satisfied when she's done her best, regardless of whether she's done better or worse than others.
- Show your daughter you aren't perfect, either. When you make a mistake, don't hide it. Then show her the world didn't end just because you messed up.
- Humor always helps. Laughing at your own mistakes will encourage your daughter to see that it's okay to laugh at hers. A bit of humor and perspective helps puncture the perfectionist urge.

- Look back together at failures that no longer sting, or obstacles she's overcome. It's a useful way to encourage perspective and resilience.

Push Out Pink

Lego, the famous toy maker, had a breakthrough idea in 2011: Bring out a line of pink blocks and sell them in princess sets. It was a move that both pandered to stereotypes and was brilliant business. The company tripled the number of little girls buying their blocks and significantly closed the Lego gender gap.

When you're up against billions of pastel dollars that color girls into a separate box, not to mention your own unconscious prejudices, reversing these cultural currents can be all the more challenging. But if girls are going to get the confidence their boy playmates seem to come by so handily, we need to break the stereotype and show our daughters they can just as well be engineers, tech whizzes, and financial geniuses.

Take the example of science and math. A 2009 report by the Organisation for Economic Co-operation and Development, the club of most developed countries in the world, showed that less than 5 percent of girls expect to work in engineering and computer science when they grow up. The figure for boys is 18 percent. Since both of these fields rely on strong math skills, the assumption might be that girls aren't doing so well in math. But that's not so. The report confirms that in problem solving there's almost no difference between the grades of boys and girls. Indeed, in some countries (Iceland, Norway, and Sweden), girls did better than boys, and only in Macau, the administrative region of China, did boys do better. Girls are perfectly capable at math—they just think they aren't. Indeed, the OECD report confirms everything we've found out about girls and competence and confidence. "Females tended to report lower mathematics-related self-efficacy than males in almost all countries," the report said, "while males tended to have a more positive view of their abilities than females. Females experienced significantly more feelings of anxiety, helplessness and stress in mathematics classes than males in 32 out of the 40 countries."

Even if your daughter likes the pink Legos or lacey ballerina dress, there's no reason you can't steer her toward math and science at the same time. We just need to change the way girls relate to them.

A few tips:

- Create an ongoing narrative for your daughter that places her in a scientific world. The weather, climate change, our food, how we travel, illnesses and allergies, the computer she Facebooks on, are all areas of science that can fire her imagination. Teachers have found that when girls get to middle school they're much more receptive to studying science when they see it presented as social studies than when they see it as a stand-alone subject.
- Don't fall into the trap of putting down your own math ability, even as a joke. How many women do you hear say, "Oh, I'm useless at math"? That self-assessment plays right into the false stereotype that girls are good at writing and boys are good at math. Instead, build up how helpful and cool you find math to be, in bite-sized ways. You are her most powerful role model.

And let your daughters get physical, even if it doesn't seem to be their natural instinct. Sports are a vital way for girls to learn how to openly compete. Karen Kelser, who runs one of the top soccer programs for young girls in Washington, DC, firmly believes that playing sports provides essential training, not for scholarship purposes, or for the Olympics, but for the real world. "It mirrors life as not much else does," she says. "There aren't that many other opportunities for girls to work as a team, to win, to lose, and to learn to get over failure, and to help each other get over failure."

She focuses the girls in her league on mastering skills rather than racking up quick wins, and though that might occasionally frustrate competitive parents, Kelser thinks loss is healthy. Moreover, she says that helping the girls build solid skills over a longer time gives them more lasting confidence.

She worries that by high school many girls stop playing competitive sports because the intense focus on winning over longer-term development pushes potential players away.

- If your daughters play a sport, don't let them quit when it gets hard—no one in sports is perfect.
- Start them young if you can. It's easier to get used to knocking into other people when you are four than it is at ten and, for girls, that can be even harder to get used to.
- Even if your daughter isn't inclined to be part of the rough-and-tumble world of soccer or basketball, think about swimming or karate or track. The daughter of a friend of ours doesn't like team sports, but just started squash at twelve and loves it. Let them excel and fail at something besides their homework or an exam.

And pointing out role models in all of these fields, whether it's science, business, politics, the arts, or competitive sports, is essential.

Role models open up a window onto the possible and encourage our daughters to push themselves to reach a tangible goal represented by a human, female face, rather than aspire to a dubious fantasy, embodied by a satin-draped, tiara-crowned piece of plastic.

Claire felt the power of role models in a profound fashion when we were researching this book. She took Della along with her to watch the Mystics basketball team practice. Della is an avid basketball player, and she was dressed accordingly, even carting her own ball. She'd seen grown-up women basketball players only on television, and when she saw the Mystics in person she was stunned into silence by all that female agility and muscle and height out on the floor. A while later, Claire and Della went to the restroom. Della usually doesn't have much use for mirrors, but she stopped in front of one for a full-length gaze such as Claire had never seen. "She was evaluating herself—turning a bit—dribbling her basketball—clearly contemplating what it takes, based on what she'd just watched," Claire remembers. "*Half an hour* of exposure to those players may have opened up a mental door. With a nod at whatever it was she saw, she turned and said, 'Come on, Mom. Let's get back there.'"

Be Kind, Honest, and Firm

Many of the lessons we can pass on to our daughters apply to the other women in our lives as well. Sometimes it's enough to tell someone that confidence is a choice you can make. Just being aware of that will spur some of the women you know to grow their confidence reserves. Often, they'll need more concrete advice. So how do you get all of those friends, colleagues, and younger women who are really talented but unable to believe in themselves to trust that they can succeed?

A solid first step is to encourage other women to recognize and talk more about their successes. Research shows that not only does that help us to reframe our thinking, but it also leads directly to more workplace successes.

A 2011 Catalyst survey of three thousand MBA graduates looked at what happened when women employed the nine strategies of a mythical ideal worker—making their career goals clear, requesting high-profile assignments, and putting time into cultivating bosses, for example. Eight made no difference at all. One worked very well, though: making their achievements known to their superiors. According to Catalyst, the women who employed that strategy "advanced further, were more satisfied with their careers, and had greater compensation growth than women who were less focused on calling attention to their successes."

There's no reason young women should feel apologetic about flagging their successes. The guys at the office do it all the time. A man is usually comfortable going to his boss with a big smile, a high five and a loud boast about his awesome triumph. Let your friends or mentees know they need to tout accomplishments, and that employers want to know. They can broadcast accomplishments without sounding like self-important braggarts: "Did you hear we won the award for creative direction? I'm so proud of my team."

Former U.S. secretary of state Madeleine Albright has a saying that there is a special place in hell for women who don't help other women. Fortunately, many women in senior positions are doing all they can to bring up those who follow in their wake. They realize their own success will be measured by the talent legacy they leave.

IMF chief Christine Lagarde takes obvious pride in the confidence she's in a position to help build. "We are in a leadership position; it's our duty to the community to go and seek the contribution of women." She describes how in meetings or news conferences she will actively seek out the woman who is afraid to raise her hand. "The body language and the eye contact tells you a woman is prepared, but she just doesn't cross the line of raising her hand or making a contribution." And that's where Lagarde will jump in and call on that woman. " 'You, in the back, you want to say something? Come on, join in.' And then it's beautiful," Lagarde says, with her characteristic open smile.

Lagarde has also become fed up with men telling her how much they'd like to promote women but just can't think of any who are qualified for that top post. So she's developed the List. In her purse, she carries the names of well-qualified women who she believes would be an asset to any organization. When a man tells her he just can't find a top woman candidate, out comes The List.

Empowering endorsements from prominent women are terrific, but a more practical way to build everyday confidence is to push those around you to try something new or to aim a bit higher. Sometimes, we let our desire to be kind and supportive get in the way of being honest.

Claire has a friend who is enormously talented and who keeps talking about setting up her own business, but she never actually does. Instead, she keeps identifying hurdles: a business partner wasn't quite right, her clients would be in Europe, she doesn't want to take out the loan she'd need to get started. For several years, Claire responded with gentle sympathies, and then one day she was blunt, telling her friend that she couldn't listen to any more excuses. Claire was worried her friend would never speak to her again, but that candor turned out to be the catalyst that got her going.

It's a great thing about women that we are so supportive of each other, but sometimes the support a friend or colleague really needs is a push. When someone is feeling down or has hit a hurdle, our temptation is to sympathize. When they're feeling bad about themselves, our nurturing selves kick in with a booster phrase like, "You're great just as you are," and then we suggest they'll feel better if they repeat this mantra to themselves.

Pattie Sellers has some true friends. She didn't listen right away, but they harangued her for a few years, and bluntly, about the fact that she needed a promotion, that she was being undervalued. She'd worked for twenty-five years at *Fortune* magazine. In addition to being an editor at large, and handling many of the big interviews, she'd largely created and now oversaw *Fortune*'s Most Powerful Women meeting every year, which had become wildly successful. Pattie wanted to stretch somehow, but was afraid to suggest it to her bosses. She already had a terrific job, after all. Why rock the boat? Finally the friends got to her, and her own inner voice as well. "I was sooooooo nervous about the meeting," she remembers.

You can probably guess how this story ended. Pattie got a big promotion, a new title, and a big raise. She now directs development of all live events across Time Inc. "I don't know what I was thinking during those years (years!) it took me to get up the courage," she confesses. "Did I think I would be fired for asking to do more? I guess that's what I was afraid of." She pauses, thinking. "I guess I thought if I asked for more, they would fire me." Having spent a fair amount of time around Pattie professionally, we were almost thunderstruck to hear she had even had this dilemma. But the experience, in the end, gave her a huge confidence boost. "Ever since I asked for a bigger, broader job and got it, I have the sense that the bosses are valuing me a lot more than they used to. Lesson learned."

Just to summarize:

Reality: *Bosses clearly think employee is enormously valuable. She can easily ask for a promotion.*

In her head: *She shouldn't ask, or she might be fired.*

That the female internal self-evaluation system can be that broken is, again, astounding, shocking, stunning, and every other word we've used multiple times that implies those things. It's just nuts! And this is why friends, acquaintances, and even strangers need to become honest and pushy with each other.

All those Hallmark "you're the best" sentiments might not work so well, anyway. A few years ago, University of Waterloo psychology professor Joanne Wood conducted a study that found positive self-statements such as "I'm great, I'm perfect, and I am lovable" can actually do more harm than good. Wood and her team conducted a study in which they asked participants to answer the ten questions in the Rosenberg Self-Esteem Scale. They then separated the participants into three groups depending on how they ranked on the scale. The people who scored lowest on the Rosenberg scale were deemed low self-esteem while the highest were put in the high self-esteem group, and those in the middle were labeled medium self-esteem. The people in the lowest and the highest groups were then randomly assigned one of two tasks. They either had to continuously repeat to themselves the statement "I'm a lovable person" for four minutes, or they had to write down their thoughts and feelings for a period of four minutes. Wood's results showed that the people who'd been in the low self-esteem group and were assigned the "I'm a lovable person" mantra felt worse about themselves after repeating the phrase compared with the low self-esteem people who'd had to write down their actual thoughts and feelings. Wood believes the findings resulted from the gap between what participants were *told* to feel and what they *really* felt. Repeating empty statements served only to underscore how far they felt they were from an ideal state of mind. The whole exercise made them feel like a double failure.

So, rather than repeatedly telling your friend she's great, try encouraging her to take action instead. Often, it takes just one suggestion—one comment from a friend or coworker. "You should consider that city council seat." "I'm sure you could handle the supervisory job. You should go for it." We can help each other most by giving each other permission to act. One little nudge might be all we need.

Sometimes that nudge comes when you are least expecting it. The power of even a fleeting image of what's possible can be critical for instilling confidence. If you're a woman in a position of authority, you can bet other women and girls will be watching. You are a role model just by virtue of who you are and the position you hold. You should know that what they see in you can change their lives. It's true in the United States, and it's often even more starkly true around the world.

We were reminded of that on a visit to the State Department, where we'd been invited to speak with two hundred women from around the world, all rising leaders in their countries. We were there to speak about the increasing power of women in the workplace; what we encountered was a humbling reminder of how fortunate we have been.

These were women who had started businesses, run for parliament, and braved political oppression. One by one, they rose to stand at a microphone and talk about the life they wanted for themselves and for their countries. (They were particularly interested, and mystified, by the fact that our husbands actually supported our careers and even pitched in with housework and child care. A useful reminder, next time we're tempted to complain.)

After the group discussion we sat down with Eunice Mussa-Napolo, from Malawi, who told us about the woman who changed her life without even realizing it. Eunice grew up in a small village. She never imagined she'd go to school, let alone have a job. That's not what village girls did in her country. She saw herself getting married at twelve or thirteen, having children and working tirelessly so the boys could get an education. "Even the six-year-old girl will wake up in the morning, fetch firewood, make breakfast for the boys to go to school," she told us, without a hint of self-pity.

But, one day, she saw something radical: a woman driving a car. Eunice had never seen or imagined anything like it. Where she came from, women simply didn't drive cars. It was an inconceivable act of independence, confidence, and guts. Even as a child Eunice was bold, and she approached this mysterious woman and struck up a conversation. If the driving had been inspiring, what she heard next was downright revolutionary: The woman was a bank manager in a distant town. Eunice was awestruck and accepted the woman's one piece of advice: "The only way to do what I do," said the bank manager, "is through education." So Eunice begged and bothered her father until he finally gave in and allowed her to sit next to the boys in class. She had seen, just fleetingly, what she might achieve. But it stuck.

Initially, she followed her role model's path precisely, becoming a bank manager herself. Even that wasn't enough for this ambitious and increasingly confident young woman. "I guess I just became passionate about the plight of the girls," she told us. Eunice decided to do something truly risky: run for a seat in parliament. After a brutal campaign against nine male opponents and with no political connections of her own, she achieved the unthinkable. She won.

8

THE SCIENCE AND THE ART

We'd both received the emails, from Genomind and 23andMe, a few days apart, letting us know the results were in. They gathered digital dust in our inboxes for a while. We later laughed about the fact that we dragged our feet before summoning up the nerve to expose our genetic secrets, to the point that neither even mentioned to the other that we had emails and should get going.

Early on, when we were plowing away into all of the science, we'd thought testing our genes was such a great idea. But now, the unopened results seemed less fascinating than portentous. What if our genes suggested we had alarming weaknesses?

Finally, our sense of curiosity won out, and we scheduled phone consultations with the genetic experts at each company. The information we got from 23andMe in its emailed summary was extensive—a detailed road map to future health hazards. It covers genetic links to many major illnesses, and as we scrolled through we traced our prospects of succumbing to everything from Alzheimer's to heart disease. We had leapt somewhat blithely into the new frontier of gene testing, but when you see your potential future laid out in black and white, it's a sobering moment. We were mostly lucky; there were no life-altering surprises in our DNA. We also learned cool, if somewhat random, facts about ourselves.

Katty has genes similar to those of some of the world's greatest sprinters. (A shame she gave up running in high school.) And she apparently can't distinguish bitter tastes well. Claire's genetics confirm that the family lore about a Native American great-great-great-great-grandmother is likely accurate, and she's since had dozens of follow-up notices about potential DNA relatives.

The results of our psychological gene profiling were much more of a surprise. 23andMe was able to test the worrier/warrior or COMT gene for us. Claire had been certain Katty would be a warrior, someone who performs well under stress, while she herself would be a worrier. Emily Drabant Conley, a neuroscientist who works for the company, told us, however, that our COMT variants are *both* met/met, the scientific identifier for worriers. That was a curveball. Research suggests met/mets perform very well in routine conditions but are much less confident and calm in high-risk settings. Drabant Conley made us feel better by reminding us that strong cognitive function often goes along with being worriers.

Our oxytocin results were also unexpected. Katty was certain Claire would have the nurturing and cuddly variant of the OXTR gene and that she would not. Wrong again. We *both* have the variant that makes us prone to feeling good about people and the world. We seized on that as a welcome counterbalance to that worrier news.

We then talked to Dr. Jay Lombard at Genomind, who had promised to explain in layman's language what his findings from our saliva tests actually meant for us. Genomind, you'll remember, does similar, but often more in-depth testing, usually directly for doctors. Scientists there are able to test for the serotonin transporter gene, the one we described in chapter 3, which can be critical for confidence. If you have one or two short strands of this gene you may be prone to anxiety, and two long strands mean you have a genetic predisposition to being more resilient. We'd both hoped we were long-stranded. We wanted that safety net of resilience embedded in our DNA. (We are aware that we were just blatantly and irresponsibly simplifying the complicated subject of genetics and its implications here, because we know genes aren't determinative, but at that moment, on the phone, the genetic stakes felt quite black and white.)

As we listened to Dr. Lombard tell us again that genes don't mean everything, we had a sense he was trying to pave the way gently. Sure enough, he told us we both have a combo short/long serotonin transporter gene. Essentially, we are more prone to anxiety, and possibly depression, depending on the challenges life tosses our way. Dr. Lombard agrees with our conclusions about the serotonin transporter and COMT genes—he thinks both are critical in terms of impact on personality in general, and on the narrower trait of confidence. Indeed, when we pressed him, he said if he had been handed our data without knowing anything about us, he would have concluded that neither of us was naturally endowed with the best building blocks for confidence, although he too thought our oxytocin variant was likely a plus.

But, again, he reminded us, "It's all probability, not actuality. The environment, epigenetics, is what turns genes on or off." And he noted that, "People with the short strands are also more vigilant and adaptable, and perhaps more likely to survive dangerous situations in the long run." We may be anxious, but we'll survive. That thought offered a bit of comfort.

The way we saw it, we had two so-called bad genes for confidence and one good one. We are more likely to worry, and be anxious, but we have a natural optimism and warmth toward the world. At some moments in our lives, that does describe us, and yet, at other times, not at all.

For a few days the image of ourselves as stressed-out worriers unsettled us. Perfectionists that we are, we felt somehow that we'd failed a test and had produced weak strands. Our slightly exasperated husbands suggested maybe we shouldn't have done the tests at all if the results were just going to make us more anxious. Then we came to realize something else about our supporting role in the nature/nurture debate. We had, both of us, clearly imposed our own natures onto our genetic footprints. We'd created those side roads and byways that neurologist Laura-Ann Petitto had described to us. Although we both have the same three gene variations, we're hardly the same people personality-wise. Katty is much more decisive, while Claire is more deliberative, for example. Katty likes physical risk but is nervous about difficult confrontations. Claire has no problem picking up the phone to engage in an awkward negotiation but doesn't relish throwing herself down the steepest slope a mountain has to offer. Maybe if we'd been younger, these genetic tests would have had more of an impact on us. As it was, they ended up intriguing us, but not defining us.

Moreover, if that serotonin transporter variant that we have, with those ominous short strands, is in fact a *sensitivity* gene, as Suomi believes, that means we are both more adaptable and reactive to our environment and that we very likely had good upbringings, since we're basically happy and successful and stable. (We now have all sorts of new appreciation for our mothers, who may have done the same heroic work as Suomi's rhesus mothers. Yes, his research suggests monkey mothers, at the moment, are more critical than fathers.)

We came to realize that the detour into our own genetics is much like the larger story of confidence we uncovered in our book. We may have started with sketchy DNA. We've both had to overcome nerves and stress and anxiety in our lives. But we've learned how. Our life experiences, we believe, outweigh any genetic dictates. The result has been that today we can actually work and perform with genuine self-assurance even in conditions of high stress. (Live television is nothing if not pure, undiluted, and immediate stress.)

Our genetic results alone didn't explain us to ourselves. We had data that didn't fully correlate with who we have turned out to be. And that plunge back into the scientific mystery of it all prepared us for our final revelation about the perplexing makeup of confidence: It can't be one size fits all.

On our mission to crack the confidence code, we have seen plenty of displays of confidence—from women who wield basketballs and guns, women who rely on textbooks and test tubes, women who frequent the halls of Congress and corporate suites. As the full picture came into view, we realized something we did not expect or consider: Confidence in women often looks different than it does in men.

We're not suggesting that there's a female version of confidence, like a mommy track, that is somehow less rigorous and ultimately sidelines us from the most engaging and rewarding professional challenges. What it means to be confident—and what it *does* for us—that's the same for women as it is for men. Doing, working, deciding, and mastering are gender neutral. But we've come to see that even when confidence is fully expressed, the style and behavior of that expression does not have to be a factory-built, generic display.

It certainly does not have to be what even today remains the prevailing model of confident behavior, and is honestly not much different from what you would have seen a generation ago: a commanding (most likely) male figure showboating, acting like a decider, and asserting his authority over others.

Male workplace bravado—perhaps testosterone-fueled and *Mad Men*–inspired—is still the gold standard. It is currently the *only* standard. The drive to win no matter the cost. The boundless craving for risk. The propensity for quick decisions. The emphasis on high-decibel and high-energy interaction. Those are values and methods of behaving. Sometimes they work. But they're not the definition of confidence.

For so long that's the way we thought women had to play it if we wanted to win, and if we wanted to experience self-assurance, even if it felt forced and phony. It was as if donning the armor of a male version of confidence would somehow transform us.

Fortunately, that doesn't have to be the way it is—especially not for women (and probably not for a lot of men, either). To put it plainly, we'd finally resolved the frustrating conundrum we'd been dragging around with us for more than a year: Do you have to be a jerk to be confident? No, thankfully. Our research and conversations with dozens of high-powered, confident women point in another direction, one that feels a lot more natural and authentic. It's an approach altogether different. The ingredients are the same, but in the end, the product can be unique.

This nuance is essential to understand, because if we don't pay attention, women will surely find ourselves chasing the wrong thing, yet again. (Neither of us has fully recovered from shoulder pads and bow ties.) Peggy McIntosh, the Wellesley scholar, thinks we need to understand that confidence has simply been "socialized over the years to be more aggressive in its display, but that it's actually much broader and often more subtle."

So, what might our brand of confidence look like? For simplicity's sake, here's a way to think about it:

Imagine yourself in that modern confidence crucible—the high-powered office meeting. You've got a critical point to make about an upcoming project, and you know it won't be popular. Often, for women, this becomes an internal wrestling match. Self-doubt might prevail, leading to silence. Or we may become so doggedly intent on conveying authority and confidence, and on not falling silent, that we make our case stridently, even a little defensively, but without authenticity.

We're saying that there's a third way. We don't always have to speak first; we can listen, and incorporate what others say, and perhaps even rely on colleagues to help make our point. We can pass credit around, and we can avoid alienating potential enemies. We can speak calmly but carry a smart message. One that will be heard. Confidence, for many of us, can even be quiet. Any of that might be the way confident behavior looks for women. (Sure, aggression comes naturally to some of us, but more often than not it comes across as inauthentic, and it is hard to feel confident when you're playing a role.)

Perhaps ours is a breed of confidence that even allows for displays of vulnerability and the questioning of our decisions. Indeed, psychologists are coming to see that there may be an unexpected, untapped power in learning to express our vulnerability, and that for many, doing so can lead to more confidence.

But we need to be clear here, because more than a few people, notably our husbands, said *hey*—how can it suddenly make sense to show a weakness, or condone second-guessing yourself? Haven't you been saying all along that's *unconfident* behavior?

In order to navigate this subtle difference, we offer some examples: expressing some vulnerability can be a strength, especially when it connects you to others. Dwelling on insecurities, and basking in self-doubt is not. Reviewing your decisions with an eye to improvement is a strength, as is admitting mistakes. Ruminating for days over decisions already made or those to come has nothing to do with the confidence we envision. Our confident behavior cannot be apologetic or mumbling or retiring. Indeed, we have to be heard, and we have to act, if we want to lead. Our instincts, if we can locate them, will help us greatly. We need to start trusting our gut.

It's what we had found in a number of different women with a number of different styles: in the crisp compassion of Valerie Jarrett; in the open, inquisitive warmth of General Jessica Wright; and in the remarkable candor of Linda Hudson. Like precious stones, we gathered up bits and pieces of the conduct of the many self-assured women we interviewed—trying to resolve, over the course of our project, the confidence we had in our heads when we started out with the very different, and appealing, brand we occasionally saw on display. Decisiveness and clarity. Approachability and often humor. The qualities varied. Christine Lagarde's mix of nerves, vulnerability, and elegant self-assurance no longer appeared contradictory or puzzling. Mainly, these women just seemed comfortable with themselves.

As we were in the throes of writing this book, Katty realized that she is at her most confident as a journalist when she trusts her instincts in interviews instead of succumbing to the pressure to follow the prescribed path of being super-combative. "There's huge pressure in our field to ask the tough-minded questions with a certain bravura. You have to be seen to be aggressive, to go for the 'gotcha' question. I've always worried that I'm not very good at it, simply because it's not really who I am and it shows that I'm pretending. Then I realized that this type of interview style isn't necessary, that it is more about getting attention for the reporter than the person being interviewed. Somehow that helped me take the pressure off myself and made my questions more natural and instinctive. It's all about developing the confidence to do it my own way."

We want to pause here and make plain that we don't have blinders on. There are still plenty of old-school managers occupying corporate suites who have a more traditional idea of what confidence looks like, and it's not particularly feminine. Sometimes we may have to pound on a table or two to please a Cro-Magnon boss. We've both learned how to interrupt loud and overbearing self-proclaimed experts on television, so that we can make our points. But there is deepening evidence that a more expansive view of confidence is actually the one that's catching hold in the workplace. A recent Stanford Business School study shows that women who can combine male and female qualities do better than everyone else, even the men. How do they define the male qualities? Aggression, assertiveness, and confidence. The feminine qualities? Collaboration, process orientation, persuasion, humility.

The researchers followed 132 business school graduates for eight years and found that women who had some of the so-called masculine traits, but who tempered them with more feminine traits, were promoted 1.5 times as often as most men, twice as often as feminine men, three times as often as purely masculine women, and 1.5 times as often as purely feminine women. Oddly, the study didn't find an advantage for men who tried to straddle the traits.

The headline to take away from this research is that women should not jettison what may be natural advantages. We need to cut our own path, and, on the matter of confidence, we need to be our own role models. Macho does not have to be our mantra. U.S. Senator Kirsten Gillibrand is adamant on this point. When we asked her whether it would be better if women, like men, were able to go around saying, "I am the best" she shrugged. "Why would you want to? You don't want to turn women into men. You want to make women celebrate their own strong points. They just need to recognize they are not deficient in any way. They just need to know what it takes to be successful and define that in a way they fully understand."

Gillibrand is fed up with what she considers the bogus assumption that, in the Senate, he who talks the loudest or longest is somehow the most effective. Her sentiments are reinforced by the findings of a recent Stanford University study: Female members of Congress get significantly more legislation passed than do the men, and work more often with members of the other party. (Maybe that's all happening while the men are pontificating on the House or Senate floor.)

Michael Nannes, chariman of the national law firm Dickstein Shapiro and another of our generous male sounding boards, firmly believes that confidence can be displayed in many forms and reports that he himself favors a less aggressive style. Nannes suggests women look for a surgical opening when trying to work their way into a male-dominated conversation. "Make a point of having a different point of view," he says. "Speak with authority, and be remembered for making a contribution."

It's worth pointing out that we often have different values in the workplace, which can affect the way we project confidence. Women, for example, according to years of corporate research, tend to have other priorities beyond profit and earnings and their own place in a hierarchy. They are often more concerned with the morale of workers and the company's mission, for example. Imagine—if you are trying to build your own stature, you're more likely to try to dominate conversations. If your goal, however, is to build consensus, you will listen to other people's opinions.

Confidence, and success, comes from playing to your distinctive strengths and values. That notion has become a popular leadership development tool. Ryan Niemiec runs the education program at the Values in Action Institute (VIA), the leading organization in the United States for the study of character strengths. "It's transformative for people to actually focus on what their strengths are," he says. "Most of us have a kind of strength blindness."

Niemiec says knowing your innate strengths doesn't mean that you never focus on areas of improvement. For example, if he works with a woman struggling with confidence as she moves up the corporate ranks, he would certainly advise her to first focus on her strengths and on making the most of them. However, she might also clearly benefit from developing more persistence, even if it's not one of her natural strengths.

Claire was surprised when she took the online VIA survey—not so much by the results, but by the interpretation. "I knew I had a good knack for people, but I'd never thought of it as my top character strength. But social intelligence came out on top for me. I learned I'm emotionally intelligent and that I'm adaptable in various social settings, and I connect with others. All true. That's probably why I like reporting so much. But *knowing* that is a strength—that it is valuable in the workplace—that would have made me much more confident ten years ago."

This, then, is the art of confidence. It is how we each create a confident interaction with the world that builds on who we are as women and as individuals. It is how we marry listening to the opinions of others without apologizing for our own. It is how we ask for votes, or donations, or support, while taming the voice that tells us we are being self-aggrandizing. It is how we reconcile what we both thought for more than a year was an irreconcilable paradox: our reluctance to speak up and take center stage, with the absolute need to make our voices heard.

Dare the Difference

Authenticity. That's what we're driving at here. It was the last part of the code to come to us, but it may be the linchpin. When confidence emanates from our core, we are at our most powerful.

In retrospect, we realized that was exactly what IMF head Christine Lagarde was talking about when she warned us, at our dinner, that leaning in the same way that the men do might force us to sacrifice what makes us unique. And, on that women-run Davos panel she had told us about, the women were actually displaying authentic confidence, as they listened, and took turns, and it looked nothing like the usual fare, which was being served up in the aggressive performance of the lone man on the panel.

Lagarde had also told us another story about making a virtue out of our differences, instead of trying to hide, erase, or change them. As we came to the end of our investigation, it seemed especially resonant. A new female president in the developing world decides she will make a change to a tradition. None of the previous presidents of her country, all of them men, would leave the palace without a suite of twenty-five cars in their entourage. But this new president finds that unnecessary. The country is broke. She decides she'll use five.

"She goes forth," says Lagarde, "but people tell her, especially the women, 'Why do you do that? You're doing it because you're a woman, and, therefore, you're going to undermine the status of the president as a woman, and people will then know that a woman is less than a man.'"

Lagarde, who serves as an unofficial counselor to female leaders around the world, didn't hesitate to offer an opinion. "I told her to dare the difference," she continues. "Make it your selling point. Don't try to measure yourself, your performance, your popularity, against the standards and the yardsticks and the measurements that men have used before you. Because you start from a different perspective, you have a different platform, you want to push different initiatives, and you should be authentic about it. So she stuck with the five. But it's hard. I can't imagine the pressure. And I'm making a point of calling her every month now to say don't give up."

Dare the difference. That we like. "You have to be savvy about it," Lagarde allows, "but you also, in a sense, have to be confident about the difference."

Two years ago, when we embarked on this project, the "problem" of our lack of confidence loomed large thematically. We envisioned The Confidence Gap as a title, in fact. As journalists, we were exhilarated by the puzzle; as women, we were gloomy. Our early research churned out stories and statistics that seemed hard to battle, an outlook that could take generations to shift. Occasionally we'd wonder whether women were destined, somehow, to feel less self-assured.

But as we deconstructed confidence and picked painstakingly through the scientific and the social and the practical findings, we started to see some glimmers. Suddenly, our confidence rush was on, and we grabbed our pans with a prospector's fervor, sifting away dirt and sand, swirling the remains until we found plenty of nuggets that had been overlooked, unexamined, or simply unearthed. We tested them and prodded them and ran them through our gauntlet of experts and researchers, until we were certain which rocks were pay dirt. Those became our path to creating confidence—our code—and we've boiled it down to the very basics:

Think Less. Take Action. Be Authentic

Confidence *is* within reach. The experience of it can be addictive. And its greatest rewards aren't fully characterized by workplace achievements or outward success. "I feel fully engaged and connected and a little high, like I'm accomplishing something great, and lost in the action," Patti Solis Doyle says, eyes closed, summoning a memory. "I feel rewarded," Caroline Miller tells us. "And accepted—that there's a place for me in the world, that I can achieve, that I have a sense of purpose. The Japanese word for purpose literally translates as 'that which I wake up for.' I think that's it."

We're back at the Verizon Center to get a final glimpse of Mystics basketball stars Monique Currie and Crystal Langhorne, this time in a game setting. It's a raucous crowd—the Mystics are about to make the playoffs.

Something our husbands have said to us over the years suddenly rings true: Sports *is* a metaphor for life. Because we can see it all on the court. Preparation and practice melded with a sense of purpose—the zone of confidence.

Just as the metaphor starts to strain, Crystal misses a long shot. But then, a few minutes later, she grabs a rebound, surges back to the basket, twirls to the right, charges the board with a left-handed layup, and sinks the ball. She's all power at that moment, superhuman, returning a purposeful high five from a teammate with a look that says, "I knew I could do it."

It looks easy now, but we know, and she knows, how she got there: the hours of practice that went into that shot, and into creating that state of mind and action.

Yes, she may have doubts, here and there. But she's overcome them enough to take action. She's earned her confidence. What we've just seen is extraordinary, really, better than superhuman, better even than superwoman. Because it's real, and it's attainable.

Acknowledgments

A few years ago, the two of us found ourselves at the U.S. Capitol, listening to a series of speakers talk about the ever-vexing issue of women and work. We'd just finished writing *Womenomics* and were now on a master list of some sort for every women-themed meeting. This particular session was in a crowded basement room. Meetings about women's empowerment are apparently still best conducted in nooks and crannies and with a bit of discomfort.

Much of what we heard we knew already: Women help the bottom line of companies, companies want female talent, but somehow the pipeline to the top is still broken. We listened to a litany of predictable-sounding solutions having to do with flexible hours, legislation, and meetings with mentors.

Then Marie Wilson started talking. Almost seventy, and a storied feminist and fighter on behalf of women in politics, she is titanium determination coated with old-fashioned grace. She said, as we noted in the introduction, that we should think about the challenge this way:

"When a man, imaging his future career, looks in the mirror, he sees a senator staring back. A woman would never be so presumptuous. She needs a push to see that image."

That hit us with a clarity we hadn't heard before. It immediately helped us to understand what we'd been seeing in our reporting. In fact, we told her afterward that we think women often can't even see who they already are—what they've already accomplished.

We knew this was a phenomenon we had to explore. We are ever grateful to Marie for giving us the original jolt, and to all her kindred spirits who have labored for decades on these issues to give us the opportunities we have today.

As we have learned, inspiration is just one element paving the road to action. This book would also have never made it out of our ruminative, multitasking brains without the ever-incisive thinking and encouragement that our agent Rafe Sagalyn always provides. He simply refused to accept that we didn't have another book in us. We mean it as a sincere compliment when we say he can think remarkably like a woman.

Our brilliant editor, Hollis Heimbouch, believed in the project from our very first phone call, and, as she did with *Womenomics*, layered our work with her enthusiasm and smart sensibility, helping us make the subject clear and urging us to fully use our voices. She laughed and groaned and marveled at every bit of the topic right along with us, without ever letting us lower our standards. Very much the ideal friend prototype we mention.

This book would never have made it to the presses without the meticulous attention to detail and infinite patience of Associate Editor Colleen Lawrie. Leslie Cohen and Stephanie Cooper—thanks for all of your hard work and enthusiasm as well. We are so grateful to the whole HarperCollins team.

We are eternally grateful for the help of dozens of generous and forgiving academics and scientists. They explained in painstaking detail how the human mind works, cognitively, biologically, genetically, and philosophically. They graciously managed to contain any exasperation they may have felt about conducting crash courses in neuroscience and psychology. We sincerely hope we have done them justice. Laura-Ann Petitto was such an enthusiastic guide through her work and her lab at Gallaudet and gave us hours of her time detailing the frontiers of neuroscience. Steve Suomi and his monkeys feel like pals at this point. Adam Kepecs gave us a new appreciation for rats, and very generously read the chapters, and conducted extensive email seminars late into the evening, complete with scanned drawings, on the nature of confidence. Jay Lombard and Nancy Grden at Genomind were so generous not only to offer to test the genes of two worried and warriorlike writers, but then to also spend hours on the phone talking us through the results, and the rest of our science. Fernando Miranda, as always, has been a wonderful friend. Catherine Afarian and Emily Drabant Conley at 23andMe were equally kind in doing our tests at lightning speed, and then walking us through all of their fantastic data. Tom Jessell at Columbia University was a treat to meet—a mind-brain mogul whose enthusiasm is catching. And thanks, too, to Dr. Daniel Amen, whose book on gender brain differences is impossible to put down. Daphna Shohamy, Sarah Shomstein, Rebecca Elliott, and Frances Champagne, we so appreciate your insights on the latest in brain science.

We were lucky, too, to be helped by some of the best academic psychologists in the world. They took us through the complications of confidence and disabused us, with great patience, of many of our preconceptions. Our holistic understanding of this mysterious and powerful quality is also thanks to them. Richard Petty was the model of calm, fielding every question, and painstakingly distilling the many shades of confidence into a handy version that we could grasp. Cameron Anderson wowed us with his research on the power of confidence over competence. Zach Estes showed us that there is a gender gap, but that it is only in confidence not ability. Women can indeed park cars as well as men. Peggy McIntosh and Joyce Ehrlinger both always seemed genuinely happy to talk with us—again. Jenny Crocker, Carole Dweck, David Dunning, Victoria Brescoll, Brenda Major, Christy Glass, Kristin Neff, Nancy Delston, Ken DeMarree, Shelley Taylor, Suzanne Segerstrom, Nansook Park, and Barbara Tannenbaum were all thoughtful and thought-provoking instructors. Ryan Niemiec taught us the importance of values in our equation. And we are heartbroken that we won't get to meet Chris Peterson. Our interview with him was so clarifying and interesting. His remarkable spirit was plain, even in a thirty-minute call.

Sharon Salzberg provided a peaceful and insightful interlude. The powerhouse women at Running Start, Susannah Wellford Shakow, Katie Shorey and Jessica Grounds (who has since gone on to try to get Hillary Clinton elected), thank you for your invaluable help. What an important mission Running Start has.

Monique Currie and Crystal Langhorne not only played amazing basketball and let Della shoot a few hoops, but they also offered an unusual look at the scale of the confidence gap, even in their imposing arena. Coaches Mike Thibault and Karen Kelser, thanks for your time and everything you do for women and girls. Michaela Bilotta, we will be ever grateful for your candor in sharing hair-raising anecdotes from Annapolis. You DO deserve everything. You earned it. Chrissellene Petropoulos, you made us laugh and appreciate the hazards of a life on a real stage. Eunice Mussa-Napolo, your story is one we will never forget.

We talked to quite a few public figures, women with busy lives and endless obligations, and yet who were eager to help all of us solve the confidence equation. Christine Lagarde, Senator Gillibrand, Secretary Chao, Valerie Jarrett, Linda Hudson, General Wright, Jane Wurwand, Clara Shih, Michelle Rhee—thank you. And a few special words of gratitude to Sheryl Sandberg, who years ago offered two relative strangers early enthusiasm, valuable direction, and a mandate to challenge our assumptions. She then very generously and rather unexpectedly read our almost-completed manuscript in the middle of her winter holidays, offering quite concrete and incredibly useful suggestions. How terrific it was to experience firsthand what we'd been told by the experts—that the truest support one woman can offer another isn't necessarily comfort or commiseration, but rather the power of her attention, her thoughts, and her honesty. Thank you, Sheryl.

And to our friends who gave up their equally precious time and privacy and wisdom to help us with the project—Patti Solis Doyle, Tia Cudahy, Virginia Shore, Beth Wilkinson, and Pattie Sellers and Tanya Coke—you were candid, funny, and inspiring as you laid bare your horror stories. We could not have cracked that stubborn code without you, and the struggle would have been a lot less entertaining.

Elizabeth Spayd offered us critical vision and genuine passion about our project. We're in your debt. And Marcia Kramer, you were a godsend in your enthusiasm for helping us get every detail just right.

John Boulin, Vivien Caetano, Jonathan Csapo, and Lizette Baghdadi all helped invaluably at various stages with research, copyediting, transcribing, footnoting. Their level of commitment was a blessing, and their enthusiasm so welcome.

Our employers at the BBC and ABC have always supported our endeavors and we thank them for tolerating our temporary distraction from duty. Special thanks to Kate Farrell at BBC World News America who kept the news running flawlessly while we chased down confidence, and to Ben Sherwood at ABC who has been such a generous champion of our journalism/book writing fusion.

As is ever the case, the people forced to witness us working on this project at close range bore the brunt of our hunt for confidence. They endured our unpredictable schedules and unreliable parenting.

Katty counts her blessings every day that she has Awa M'Bow in her family—Awa's kindness and generosity are an example to us all. Claire could not have written this without the rock-solid support of Janet Sanderson, who has become part of our clan, and whose big heart is an inspiration. Thank you. And Tara Mahoney, somehow you manage to keep the Shipman/Carneys on track and laughing at the same time. We are so grateful for your sunny skills.

Our kids put up with perpetually distracted mothers who were either locked in offices, lugging around computers and stacks of papers, or citing yet another annoying statistic for them. No, despite your constant refrain, we don't love the book more than we love you. It's not even close.

We want to thank them, deeply, for the inspiration they provide with almost every twitch. It's an awe-inspiring window on confidence to watch Hugo's visceral drive to challenge convention and authority, to perform and provoke, and then his readiness to shrug off what others might think. Thank you for your joy and your ingenuity and your hugs. I love watching you unfold into such a marvelous young adult, one who already has the plans and enthusiasm enough to furnish three lives at least. And to see Jude sway a whole class to appreciate his Scottish kilt, by sheer confidence of personality. We all love watching you operate. And to witness Felix stick, with such impressive determination, through the trials of life in a totally new culture, with little sunlight to ease the days, and come out in triumph. I miss you every day but am so happy to see you having fun. Maya's self-assurance is in full, determined bloom, and we have little doubt the world will be a better place when she rules it. What I'll do without her isn't clear. We will all miss your company. And to our younger girls: Della, how searing it was to see that the world still does not look right from the point of view of a four-year-old, who lamented for more than two years, with voluble anguish, that she could never grow up to be, as a girl, her beloved Batman. You and your passions, Della, have taught me more than I ever knew there was to learn. I'm so grateful for you and your love. I wish I had half of your bravery and sense of adventure. The free world, as my friend Patti always reminds me, is waiting for you to grow up and run it. The summer Poppy first dyed her hair blue she was five, we have since gone through purple and green; you need a fairly robust sense of who you are to experiment with hair color in kindergarten. Even after all this research, I'm still not sure where you get it, but you are my confidence lodestar.

And thanks are due especially to our long-suffering, fantastically supportive husbands, Tom and Jay. They both read this manuscript with great interest, and from commas to insightful commentary, they made this a much better book. They brought us tea (Katty) and ice cream (Claire), ferried kids, and filled in ever-shifting gaps. We are very lucky. And confident that we've done well in choosing our husbands.

Finally, we would both like to thank fate for bringing us together. Or, perhaps we need to give ourselves more credit than that. Having learned the lesson of this book, we will not merely give luck all of the praise for making us friends, partners, and confidence collaborators. We thank each other for the mutual efforts we make to nurture our relationship. There are not many people in the world you can write not one but two books with, and still be the best of friends.

Notes

1: It's Not Enough to Be Good

8 **first Chinese-American cabinet secretary:** http://elainelchao.com/biography.

11 **it eventually caught on across the Atlantic:** "BBC News Profile: Dominique Strauss-Kahn," last updated December 10, 2012, http://www.bbc.co.uk/news/world-europe-13405268; "Strauss-Kahn Resigns From IMF; Lawyers to Seek Bail on Rape Charges," last updated May 19, 2011, http://abcnews.go.com/US/dominique-strauss-khan-resigns-lawyers-return-court-seeking/story?id=13636051.

14 **how confident they feel in their professions:** Jill Flynn, Kathryn Heath, and Mary Davis Holt, "Four ways women stunt their careers unintentionally," *Harvard Business Review* 20 (2011).

14 **Linda Babcock:** Linda Babcock, "Nice Girls Don't Ask," *Harvard Business Review*, 2013.

14 **problem stems from lack of confidence:** Marilyn J. Davidson and Ronald J. Burke, *Women in Management Worldwide* (Aldershot: Ashgate, 2004), 102.

15 **Think about it for a minute:** Justin Kruger and David Dunning,

"Unskilled and Unaware of It: How Difficulties in Recognizing One's Own Incompetence Lead to Inflated Assessments," *Journal of Personality and Social Psychology,* Vol 77 (6), Dec 1999: 1121–34, doi:10.1037/0022-3514.77.6.1121.

15 **notions about their ability:** David Dunning, Kerri Johnson, Joyce Ehrlinger, and Justin Kruger, "Why People Fail to Recognize Their Own Incompetence," *Current Directions in Psychological Science*, no. 3 (2003): 83–87.

16 **when in the minority:** Christopher F. Karpowitz, Tali Mendelberg, and Lee Shaker, "Gender inequality in deliberative participation," *American Political Science Review* 106, no. 3 (2012): 533–47.

16 **self-perception decades ago:** Toni Schmader and Brenda Major, "The impact of ingroup vs. outgroup performance on personal values," *Journal of Experimental Social Psychology* 35, no. 1 (1999): 47–67.

19 **30 percent better than it is:** Ernesto Reuben, Columbia University Business School Journal, Ideas At Work: "Confidence Game," last modified November 22, 2011, https://www4.gsb.columbia.edu/ideasatwork/feature/7224716/Confidence Game.

20 **play into different stereotypes:** Nalini Ambady, Margaret Shih, Amy Kim, and Todd L. Pittinsky, "Stereotype susceptibility in children: Effects of identity activation on quantitative performance," *Psychological Science* 12, no. 5 (2001): 385–90.

21 **Hewlett-Packard conducted a study:** Hau L. Lee and Corey Billington, "The evolution of supply-chain-management models and practice at Hewlett-Packard," *Interfaces* 25, no. 5 (1995): 42–63.

21 **value of confidence and competence:** Cameron Anderson, Sebastien Brion, Don A. Moore, and Jessica A. Kennedy, "A status-enhancement account of overconfidence" (2012), http://haas.berkeley.edu/faculty/papers/anderson/status%20enhancement%20account%20of%20overconfidence.pdf.

24 **described by psychologist Mihaly Csikszentmihalyi as perfect concentration:** Mihaly Csikszentmihalyi, *Flow* (New York: HarperCollins, 1991).

2: Do More, Think Less

27 **His groundbreaking studies:** Adam Kepecs, Naoshige Uchida, Hatim A. Zariwala, and Zachary F. Mainen, "Neural correlates, computation and behavioural impact of decision confidence," *Nature* 455, no. 7210 (2008): 227–31.

41 **optimism is the key to life:** Martin E. Seligman, *Learned Optimism: How to Change Your Mind and Your Life* (New York: Random House Digital, 2011).

42 **Morris Rosenberg came up with a basic self-esteem scale:** Morris Rosenberg, *Conceiving the Self* (New York: Basic Books, 1979).

To check your self-esteem level, answer the following questions with one of the four statements:

Strongly Agree
Agree
Disagree
Strongly Disagree.

1. I feel that I am a person of worth, at least on an equal plane with others.
2. I feel that I have a number of good qualities.
3. All in all, I am inclined to feel that I am a failure.
4. I am able to do things as well as most other people.
5. I feel I do not have much to be proud of.
6. I take a positive attitude toward myself.
7. On the whole, I am satisfied with myself.
8. I wish I could have more respect for myself.
9. I certainly feel useless at times.
10. At times I think I am no good at all.

You then calculate your score as follows:

For items 1, 2, 4, 6, and 7:

Strongly agree = 3
Agree = 2
Disagree = 1
Strongly disagree = 0

For items 3, 5, 8, 9, and 10:

Strongly agree = 0
Agree = 1
Disagree = 2
Strongly disagree = 3

The scale ranges from 0 to 30. Scores between 15 and 25 are within normal range; scores below 15 suggest low self-esteem.

43 **unrealistic self-esteem:** David H. Silvera and Charles R. Seger, "Feeling good about ourselves: Unrealistic self-evaluations and their relation to self-esteem in the United States and Norway," *Journal of Cross-Cultural Psychology* 35.5 (2004): 571–85.

44 **confidence and optimism as closely related:** N. Park and C. Peterson, "Positive psychology and character strengths: Its application for strength-based school counseling," *Journal of Professional School Counseling* 12 (2008): 85–92; N. Park and C. Peterson, "Achieving and sustaining a good life," *Perspectives on Psychological Science* 4 (2009): 422–28.

45 **The work of Sharon Salzberg:** Sharon Salzberg, *The Kindness Handbook: A Practical Companion* (Boulder, CO: Sounds True, 2008).

Real Happiness: The Power of Meditation: A 28-Day Program (New York: Workman Publishing, 2011).

45 **recently pioneered as an academic pursuit:** Kristin Neff, "Self-compassion: An alternative conceptualization of a healthy attitude toward oneself," *Self and Identity* 2, no. 2 (2003): 85–101.

47 **"Theory of Behavioral Change":** Albert Bandura, "Self-efficacy: Toward a unifying theory of behavioral change," *Psychological Review* 84, no. 2 (1977): 191.

51 **Estes did a series of tests:** Zachary Estes, "Attributive and relational processes in nominal combination," *Journal of Memory and Language* 48, no. 2 (2003): 304–19.

54 **two of the most trusted confidence scales:** This first test, crafted by Professor Richard Petty of Ohio State University and Kenneth DeMarree of the University of Buffalo, was just crafted in 2013. It is simple, and it's meant to provide a clear sense of general confidence. The key is to be as honest as possible in the responses.

Respond to the statements below on a scale of 1 to 9: A score of 1 indicates that you agree very much; a score of 9 means that you disagree very much.

1. I am a good person.
2. I am a sad person.
3. I am a confident person.
4. I am an ineffective person.
5. I am a warm person.
6. I am a bad person.
7. I am a happy person.
8. I am a doubtful person.
9. I am an effective person.
10. I am a cold person.

Now, isolate your results from items 3 and 8. Reverse the score for 3; in other words, if you gave yourself a 2, turn that into an 8, and then add those two numbers.

A score of 2 means that you are as doubtful as they come, and a score of 18 would be quite confident.

The average score among young people, students at Texas Tech and Ohio State, is 13. That should give you a sense as to where you fall. Professor Petty has told us that a range of 9 to 14 is average, based on data so far. Below 9 is lower than average, and above 14 is higher than average.

For more on how the scale has been used so far, see studies such as K. G. DeMarree, C. Davenport, P. Briñol, and R. E. Petty, "The Role of Self-Confidence in Persuasion: A Multi-Process Examination," presented at the annual meeting of the Midwestern Psychological Association, Chicago, Ill., May 2012; and R. E. Petty, K. G. DeMarree, and P. Briñol, "Individual Differences in the Use of Mental Contents," presented at the annual meeting of the Society of Experimental Social Psychology, Berkeley, Calif., September 2013.

The other relevant measure is the General Self-Efficacy Scale, developed in 1981 and still in wide use today. It is more a measure of a sense of belief in an ability to get things done than an assessment of a generalized sense of confidence. (English version by Ralf Schwarzer and Matthias Jerusalem, 1995.)

Answer the questions on the next page using the following response format:

1 = Not at all true
2 = Hardly true
3 = Moderately true
4 = Exactly true

1. I can always manage to solve difficult problems if I try hard enough.
2. If someone opposes me, I can find the means and ways to get what I want.
3. It is easy for me to stick to my aims and accomplish my goals.
4. I am confident that I could deal efficiently with unexpected events.
5. Thanks to my resourcefulness, I know how to handle unforeseen situations.
6. I can solve most problems if I invest the necessary effort.
7. I can remain calm when facing difficulties because I can rely on my coping abilities.
8. When I am confronted with a problem, I can usually find several solutions.
9. If I am in trouble, I can usually think of a solution.
10. I can usually handle whatever comes my way.

Your total will be between 10 and 40, with higher scores meaning a more confident attitude. A score of 29 has been about average worldwide.

3: Wired for Confidence

58 **essential to personality formation:** Stephen J. Suomi et al., "Cognitive impact of genetic variation of the serotonin transporter in primates is associated with differences in brain morphology rather than serotonin neurotransmission," *Molecular Psychiatry* 15, no. 5 (2009): 512–22.

58 **a key criterion for confidence:** Klaus-Peter Lesch et al., "Association of anxiety-related traits with a polymorphism in the serotonin

transporter gene regulatory region," *Science* 274, no. 5292 (1996): 1527–31.

58 **linked with happiness and optimism:** A. Graff-Guerrero, C. De la Fuente-Sandoval, B. Camarena, D. Gomez-Martin, R. Apiquian, A. Fresan, A. Aguilar et al., "Frontal and limbic metabolic differences in subjects selected according to genetic variation of the SLC6A4 gene polymorphism," *Neuroimage* 25, no. 4 (2005): 1197–1204.

61 **sets of twins in Britain:** Alexandra Trouton, Frank M. Spinath, and Robert Plomin, "Twins early development study (TEDS): A multivariate, longitudinal genetic investigation of language, cognition and behavior problems in childhood," *Twin Research* 5, no. 5 (2002): 444–48.

62 **link between genes and IQ:** Robert Plomin et al., "DNA markers associated with high versus low IQ: The IQ quantitative trait loci (QTL) project," *Behavior Genetics* 24, no. 2 (1994): 107–18.

62 **proclivity to be a professional dancer:** Can you truly thank your genes for your fox-trotting ability? Israeli researchers have found that professional dancers often share two gene variants—one regulates serotonin, another affects the hormone vasopressin, which in turn affects bonding and social communication. The thinking (though it's not widely accepted) is that these gene variants encourage an impulse toward creativity, communication, bonding, and even spirituality that makes dancing more appealing. Interestingly, the dancers don't share a genetic variation many top athletes have. (That, if you are wondering, is a variant of the AGT gene, which seems to be common among athletes involved in power sports, and may, according to researchers, affect muscle strength.) No word yet on a bowling gene. R. Bachner-Melman et al., "AVPR1a and SLC6A4 Gene Polymorphisms Are Associated with Creative Dance Performance," *Public Library of Science Genetics* (2005), e42. doi:10.1371/journal.pgen.0010042; Christian

Kandler, "The Genetic Links Between the Big Five Personality Traits and General Interest Domains," *Personality and Social Psychology Bulletin* (2011).

63 **comparing DNA and IQ scores:** Robert Plomin and Frank M. Spinath, "Intelligence: Genetics, Genes, and Genomics," *Journal of Personality and Social Psychology*, no. 1 (2004): 112–29, http://webspace.pugetsound.edu/facultypages/cjones/chidev/Paper/Articles/Plomin-IQ.pdf; B. Devlin, Michael Daniels, and Kathryn Roeder "The heritability of IQ," *Nature* 388 (1997): 468–71.

63 **from the world's smartest people:** John Bohannon, "Why Are Some People So Smart? The Answer Could Spawn a Generation of Superbabies," *Wired*, July 16, 2013.

65 **and to be faithful:** Navneet Magon and Sanjay Kalra, "The orgasmic history of oxytocin: Love, lust, and labor," *Indian Journal of Endocrinology and Metabolism* 15 (2011): S156.

65 **it even encourages monogamy:** Researchers in Germany rounded up fifty-seven men. Some were in committed relationships; others were single. All received a few hits of oxytocin via nasal spray before being "exposed" to an extremely attractive female researcher. As the woman moved toward and away from each of the men, they were asked to report when she was the "ideal distance" away from them. The ones who'd had some oxytocin, and who were in monogamous relationships, wanted to keep the scientific siren farther away—on average they kept her four to six inches farther away than the single men did. Now, the study didn't say just how close the single guys wanted her to stand, or what she was wearing, clearly critical details. But check out the November, 2012, issue of the *Journal of Neuroscience*—it's got a new, racy feel to it.

65 **the delivery of oxytocin:** Shelley E. Taylor et al., "Gene-Culture Interaction Oxytocin Receptor Polymorphism (OXTR) and Emotion Regulation," *Social Psychological and Personality Science* 2, no. 6 (2011): 665–72.

66 **encourages dramatic risk-taking:** Cynthia J. Thomson, "Seeking sensations through sport: An interdisciplinary investigation of personality and genetics associated with high-risk sport" (2013).

66 **programmed to worry or fight:** Anil K. Malhotra et al., "A functional polymorphism in the COMT gene and performance on a test of prefrontal cognition," *American Journal of Psychiatry* 159, no. 4 (2002): 652–54.

71 **environmental influence than others:** Nessa Carey, *The Epigenetics Revolution: How Modern Biology Is Rewriting Our Understanding of Genetics, Disease, and Inheritance* (New York: Columbia University Press, 2012); Richard C. Francis, *Epigenetics: How Environment Shapes Our Genes* (New York: Norton, 2012); Paul Tough, "The Poverty Clinic: Can a Stressful Childhood Make You a Sick Adult?" *New Yorker,* March 21, 2011; Stacy Drury, "Telomere Length and Early Severe Social Deprivation: Linking Early Adversity and Cellular Aging," *Molecular Psychiatry* 17 (2012): 719–27; Jonathan D. Rockoff, "Nature vs. Nurture: New Science Stirs Debate: How Behavior is Shaped; Who's an Orchid, Who's a Dandelion," *Wall Street Journal*, September 16, 2013.

74 **and their future offspring:** Frances Champagne and Michael J. Meaney, "Like mother, like daughter: Evidence for non-genomic transmission of parental behavior and stress responsivity," *Progress in Brain Research* 133 (2001): 287–302.

74 **through their DNA:** Rachel Yehuda et al., "Transgenerational effects of posttraumatic stress disorder in babies of mothers exposed to the World Trade Center attacks during pregnancy," *Journal of Clinical Endocrinology & Metabolism* 90, no. 7 (2005): 4115–18.

74 **more likely to be obese:** Robert A. Waterland and Randy L. Jirtle, "Transposable Elements: Targets for Early Nutritional Effects on Epigenetic Gene Regulation," *Molecular and Cellular Biology* 23 (2003): 5293–3000.

75 **thrive in many environments:** Bruce J. Ellis and W. Thomas

Boyce, "Biological sensitivity to context," *Current Directions in Psychological Science* 17, no. 3 (2008): 183–87.

76 **leaves a much larger imprint:** Avshalom Caspi et al., "Genetic sensitivity to the environment: The case of the serotonin transporter gene and its implications for studying complex diseases and traits," *American Journal of Psychiatry* 167, no. 5 (2010): 509.

76 **more able to adapt:** E. Fox et al., "The Serotonin Transporter Gene Alters Sensitivity to Attention Bias Modification: Evidence for a Plasticity Gene," *Biological Psychiatry* 70 (2011): 1049–54.

78 **part of our hard-wiring:** Richard J. Davidson and Bruce S. McEwen, "Social Influences on Neuroplasticity: Stress and Interventions to Promote Well-Being," *Nature Neuroscience* 15 (2012): 689–95; Peter S. Eriksson et al., "Neurogenesis in the Adult Human Hippocampus," *Nature Medicine* 4 (1998): 1313–17; Elizabeth Gould et al., "Neurogenesis in the Dentate Gyrus of the Adult Tree Shrew Is Regulated by Psychosocial Stress and NMDA Receptor Activation," *Journal of Neuroscience* 17 (1997): 2492–98; Gerd Kempermann and Fred H. Gage, "New Nerve Cells for the Adult Human Brain," *Scientific American* 280 (1999): 48–53; Jack M. Parent et al., "Dentate Granule Cell Neurogenesis Is Increased by Seizures and Contributes to Aberrant Network Reorganization in the Adult Rat Hippocampus," *Journal of Neuroscience* 17 (1997): 3727–38.

79 **eight weeks of meditation:** Yi-Yuan Tang et al., "Mechanisms of White Matter Changes Induced by Meditation," *Proceedings of the National Academy of Sciences* (2012): doi:10.1073/pnas.1207817109; B. K. Hölzel et al., "Mindfulness Practice Leads to Increases in Regional Brain Gray Matter Density," *Psychiatry Research: Neuroimaging* 191 (2011): 36–43.

79 **shrank and remained smaller:** B. K. Hölzel et al., "Stress Reduction Correlates with Structural Changes in the Amygdala," *Social Cognitive and Affective Neuroscience* 5 (2010): 11–17.

79 **back to the frontal cortex:** R. A. Bryant et al., "Amygdala and

ventral anterior cingulate activation predicts treatment response to cognitive behaviour therapy for post-traumatic stress disorder." *Psychological Medicine* 38, no. 4 (2008): 555–62.

79 **twelve adults with arachnophobia:** Marla Paul, Northwestern University, "Touching Tarantulas: People with spider phobia handle tarantulas and have lasting changes in fear response," last modified May 21, 2012.

81 **a series of video games:** G. Elliott Wimmer and Daphna Shohamy, "Preference by association: how memory mechanisms in the hippocampus bias decisions," *Science* 338, no. 6104 (2012): 270–73.

4: "Dumb Ugly Bitches" and Other Reasons Women Have Less Confidence

87 **by every measure of profitability:** Joanna Barsh and Lareina Yee, "Unlocking the full potential of women in the US economy," 2011, http://www.mckinsey.com/client_service/organization/latest_thinking/unlocking_the_full_potential; International Monetary Fund, "Women, Work and the Economy," September 2013, http://www.imf.org/external/pubs/ft/sdn/2013/sdn1310.pdf; Catalyst, "The Bottom Line—Corporate performance and women's representation on boards," 2007, http://www.catalyst.org/knowledge/bottom-line-corporate-performance-and-womens-representation-boards; McKinsey and Company, "Women Matter: Gender diversity, a corporate performance driver," 2007, http://www.mckinsey.com/features/women_matter; Roy D. Adler, Ph.D., Pepperdine University, "Women in the executive suite correlate to high profits," 2009, http://www.w2t.se/se/filer/adler_web.pdf; David Ross, Columbia Business School, "When women rank high, firms profit," 2008, http://www8.gsb.columbia.edu/ideas-at-work/publication/560/when-women-rank-high-firms-profit; Ernst and Young, "High achievers: Recognizing the power of women to spur business and economic growth," 2013, http://www.ey.com/Publication/vwLUAssets

/Growing_Beyond_-_High_Achievers/$FILE/High%20achievers %20-%20Growing%20Beyond.pdf.

87 **women began getting hired:** Claudia Goldin and Cecilia Rouse, "Orchestrating Impartiality: The Impact of 'Blind' Auditions on Female Musicians, no. w5903, National Bureau of Economic Research, 1997.

88 **Research shows that:** Carol Dweck, *Mindset: The New Psychology of Success* (New York: Random House Digital, 2006).

90 **in male-dominated industries:** Erin Irick, "NCAA Sports Sponsorship and Participation Rates Report: 1981–1982—2010–2011)," Indianapolis, IN, National Collegiate Athletics Association, 69.

90 **rose sixfold from 1972 to 2011:** Alana Glass, *Forbes*, "Title IX At 40: Where Would Women Be Without Sports?," last modified May 23, 2012, http://www.forbes.com/sites/sportsmoney/2012/05/23/title-ix-at-40-where-would-women-be-without-sports/2/.

91 **The Centers for Disease Control:** Brooke de Lench, *Home Team Advantage: The Critical Role of Mothers in Youth Sports* (New York: HarperCollins, 2006).

91 **Academics confirm what we know:** L. Moldando, University of South Florida, "Impact of early adolescent anxiety disorders on self-esteem development from adolescence to young adulthood," August 2013.

91 **Psychologists believe:** Carol J. Dweck, "The Mindset of a Champion," last modified 2013, accessed October 9, 2013, http://champions.stanford.edu/perspectives/the-mindset-of-a-champion/.

95 **and become overly deferential:** Anastasia Prokos and Irene Padavic, " 'There oughtta be a law against bitches': Masculinity lessons in police academy training," G*ender, Work & Organization* 9, no. 4 (2002): 439–59.

96 **on a group of men and women:** Victoria Brescoll, "Who Takes the Floor and Why: Gender, Power, and Volubility in Organizations," *Sage Journals*, last modified March 26, 2012, http://asq.sagepub.com/content/early/2012/02/28/0001839212439994.

97 **the reality of "stereotype threat":** Joshua Aronson and Claude Steele, "Stereotype threat and the intellectual test performance of African Americans," *Journal of Personality and Social Psychology* 69 (1995).

98 **a paid maternity leave:** Lawrence M. Berger, Jennifer Hill, and Jane Waldfogel, "Maternity leave, early maternal employment and child health and development in the US," *Economic Journal* 115, no. 501 (2005): F29–F47.

98 **The latest Global Gender Gap Report:** World Economic Forum, The Global Gender Gap Report, 2013.

99 **actually think we are beautiful:** "Dove Campaign for Real Beauty," *Eating Disorders: An Encyclopedia of Causes, Treatment, and Prevention* (2013): 147.

100 **women are indeed judged more harshly at work:** Sylvia Ann Hewlett, Center for Talent Innovation Study, 2011; Mark V. Roehling, "Weight-based discrimination in employment: Psychological and legal aspects," *Personnel Psychology* 52, no. 4 (1999): 969–1016; "The Seven Ways Your Boss Is Judging Your Appearance," *Forbes*, November 2012; Lisa Quast, "Why Being Thin Can Actually Translate Into a Bigger Paycheck for Women," *Forbes*, June 6, 2011, http://www.forbes.com/sites/lisaquast/2011/06/06/can-being-thin-actually-translate-into-a-bigger-paycheck-for-women/.

102 **A study of recent graduates:** Leslie McCall, "Gender and the new inequality: Explaining the college/non-college wage gap," *American Sociological Review* (2000): 234–55.

104 **dangers of excessive rumination:** Susan Nolen-Hoeksema, Blair E. Wisco, and Sonja Lyubomirsky, "Rethinking rumination," *Perspectives on Psychological Science* 3, no. 5 (2008): 400–424.

106 **a healthy sign of resilience:** Travis J. Carter and David Dunning, "Faulty Self-Assessment: Why Evaluating One's Own Competence Is an Intrinsically Difficult Task," *Social and Personality Psychology Compass* 2, no. 1 (2008): 346–60.

107 **a largely female issue:** Robert M. Lynd-Stevenson and Christie M.

Hearne, "Perfectionism and depressive affect: The pros and cons of being a perfectionist," *Personality and Individual Differences* 26, no. 3 (1999): 549–62; Jacqueline K. Mitchelson, "Seeking the Perfect Balance: Perfectionism and Work-Family Balance," *Journal of Occupational and Organizational Psychology* 82, no. 23 (2009): 349–67.

107 **authors of *The Plateau Effect:*** Bob Sullivan and Hugh Thompson, *The Plateau Effect* (New York: Dutton Adult, 2013).

109 **It Matters Where the Matter Is:** For an overview on brain differences, we'd suggest the *The Female Brain* by Louann Brizendine or *Unleash the Power of the Female Brain* by Daniel G. Amen. Also extremely helpful is the growing body of literature by researchers such as Gert De Vries, Patrica Boyle, Richard Simmerly, Kelly Cosgrove and Larry Cahill. And finally, this comprehensive review of literature is extremely helpful: check out Glenda E. Gillies and Simon McArthur's article, "Estrogen Actions in the Brain and the Basis for Differential Action in Men and Women: A Case for Sex-Specific Medicines."

109 **identify male versus female:** Doreen Kimura, "Sex differences in the brain," *Scientific American* 267, no. 3 (1992): 118–25.

109 **relative to their body size:** C. Davison Ankney, "Sex differences in relative brain size: The mismeasure of woman, too?," *Intelligence* 16, no. 3 (1992): 329–36.

109 **on language arts:** Michael Gurian, *Boys and Girls Learn Differently! A Guide for Teachers and Parents*, rev. 10th anniversary ed., Wiley. com, 2010.

109 **One Harvard study:** Tor D. Wager, K. Luan Phan, Israel Liberzon, and Stephan F. Taylor, "Valence, gender, and lateralization of functional brain anatomy in emotion: A meta-analysis of findings from neuroimaging," *Neuroimage* 19, no. 3 (2003): 513–31.

110 **critical for integrating information:** Marjorie LeMay and Antonio Culebras, "Human brain–morphologic differences in the hemispheres demonstrable by carotid arteriography," *New England Journal of Medicine* 287, no. 4 (1972): 168–70.

110 **diffusion tensor imaging:** Jung-Lung Hsu et al., "Gender differences and age-related white matter changes of the human brain: a diffusion tensor imaging study," *Neuroimage* 39, no. 2 (2008): 566–77; J. Sacher and J. Neumann et al., "Sexual dimorphism in the human brain: Evidence from neuroimaging," *Magnetic Resonance Imaging* 3 (April 2013): 366–75, doi:10.1016/j.mri.2012.06.007; H. Takeuchi and Y. Taki et al., "White Matter Structures Associated with Emotional Intelligence: Evidence from Diffusion Tensor Imaging," *Human Brain Mapping* 5 (May 2013): 1025–34, doi:10.1002/hbm.21492.

110 **A handful of studies:** Richard Kanaan, "Gender Differences in White Matter Microstructure." *PLoS ONE*, 2012: 7(6); Richard Haier, "The Neuroanatomy of General Intelligence: Sex Matters," *Neuroimage*, volume 25, March 2005.

110 **blood flow and activity patterns:** Daniel G. Amen, *Unleash the Power of the Female Brain: Supercharging Yours for Better Health, Energy, Mood, Focus, and Sex* (New York: Random House Digital, 2013).

112 **you guessed it, men rely:** L. Cahill, University of California, Irvine, "Sex-related Differences in Amygdala Functional Connectivity during Resting Conditions," 2006.

112 **anxiety and amygdala under control:** George R. Heninger, "Serotonin, sex, and psychiatric illness," *Proceedings of the National Academy of Sciences* 94, no. 10 (1997): 4823–24.

112 **it's larger in women:** Louann Brizendine, *The Female Brain* (New York: Random House Digital, 2007).

113 **as early as twenty-six weeks:** Reuwen Achiron and Anat Achiron, "Development of the human fetal corpus callosum: A high-resolution, cross-sectional sonographic study," *Ultrasound in Obstetrics & Gynecology* 18, no. 4 (2001): 343–47.

113 **in their actual capabilities:** Alan C. Evans, "The NIH MRI study of normal brain development," *Neuroimage* 30, no. 1 (2006): 184–202.

113 **boys in spatial ability:** Angela Josette Magon, "Gender, the Brain

and Education: Do Boys and Girls Learn Differently?," PhD diss., University of Victoria, 2009.

114 **muscle size to competitive instinct:** P. Corbier, D. A. Edwards, and J. Roffi, "The neonatal testosterone surge: A comparative study," *Archives of Physiology and Biochemistry* 100, no. 2 (1992): 127–31.

114 **and cooperating with others:** There's fascinating research that found the testosterone levels of young boys, in the first six months of their lives, to be as high as that of boys going through puberty. That temporary testosterone wash may have to do with brain development, the experts speculate. W. L. Reed et al., "Physiological effects on demography: A long-term experimental study of testosterone's effects on fitness," *American Naturalist* 167, no. 5 (2006): 667–83.

114 **A number of recent studies:** Anna Dreber, "Testosterone and Financial Risk Preferences," Harvard University, 2008.

115 **thcy made riskier trades:** John M. Coates and Joe Herbert, "Endogenous steroids and financial risk taking on a London trading floor," *Proceedings of the National Academy of Sciences* 105, no. 16 (2008): 6167–72.

115 **connecting and cooperating:** Amy Cuddy, "Want to Lean In? Try a Power Pose," *Harvard Business Review*, last modified March 20, 2013, http://blogs.hbr.org/2013/03/want-to-lean-in-try-a-power-po-2/.

116 **The finding of one experiment:** Nicholas Wright, "Testosterone disrupts human collaboration by increasing egocentric choices," Royal Society of Biological Sciences, 2012.

116 **different instincts from testosterone:** Louann Brizendine, *The Female Brain*. Brizendine offers a detailed sense of the impact of estrogen on personality.

116 **in that disastrous year, 2008:** Heber Farnsworth and Jonathan Taylor, "Evidence on the compensation of portfolio managers," *Journal of Financial Research* 29, no. 3 (2006): 305–24.

117 **there is growing evidence:** Henry Mahncke, "Memory enhancement in healthy older adults using a brain plasticity-based training program," 2006, www.pnas.org/content/103/33/12523.abstract.

117 **one startling study:** Lee Gettler et al., "Longitudinal evidence that fatherhood decreases testosterone in human males," 2011, http://www.pnas.org/content/early/2011/09/02/1105403108.abstract.

5: The New Nurture

122 **can cause real problems:** R. Baumeister, "The Lowdown on High Self-Esteem,'" *Los Angeles Times* (2005).

122 **bear a lot of responsibility:** Kali H. Trzesniewski, M. Brent Donnellan, and Richard W. Robins, "Is 'Generation Me' really more narcissistic than previous generations?," *Journal of Personality* 76, no. 4 (2008): 903–18.

123 **graduated exposure to risk:** Nansook Park, "The role of subjective well-being in positive youth development," *Annals of the American Academy of Political and Social Science* 591, no. 1 (2004): 25–39.

124 **and tolerance for hardship:** Amy Chua, "Why Chinese Mothers Are Superior," *Wall Street Journal*, January 8, 2011, http://online.wsj.com/news/articles/SB1000142405274870411504576059713528698754.

126 **firstborn boys in particular, also suffer:** Andrea Ichino, Elly-Ann Lindström, and Eliana Viviano, "Hidden Consequences of a First-Born Boy for Mothers," April 2011, http://ftp.iza.org/dp5649.pdf.

128 **as a "growth mind-set":** Dweck, *Mindset.*

133 **women do to feel confident:** Jennifer Crocker, "Contingencies of self-worth: Implications for self-regulation and psychological vulnerability," *Self and Identity* 1, no. 2 (2002): 143–49.

137 **as high as 50 percent:** Caroline Adams Miller, "Five Things That Will Improve Your Life in 2013," last modified December 28, 2013, www.carolinemiller.com/five-things-that-will-improve-your-life-in-2013/.

6: Failing Fast and Other Confidence-Boosting Habits

149 **getting physical with your thoughts:** Pablo Briñol et al., "Treating Thoughts as Material Objects can Increase or Decrease Their Impact on Evaluation," January 2013, *Psychological Science* 24 (1): 41–47 (published online November 26, 2012, doi: 10.1177/0956797612449176).

152 **surprising boost of confidence:** Jennifer Crocker and Jessica Carnevale, "Self-Esteem Can Be an Ego Trap," *Scientific American*, August 9, 2013, http://www.scientificamerican.com/article.cfm?id=self-esteem-can-be-ego-trap.

163 **and an optimistic mind-set:** Zak Stambor, "A key to happiness," *Monitor on Psychology* 37, no. 9 (2006): 34.

164 **Sitting up straight will give you a short-term confidence boost:** Pablo Briñol, Richard E. Petty, and Benjamin Wagner, "Body Posture Effects on Self-Evaluation: A Self-Validation Approach," *European Journal of Social Psychology*, February 25, 2009.

7: Now, Pass It On

168 **learning has a big impact on confidence:** James W. Stigler and James Hiebert, *The Teaching Gap: Best Ideas from the World's Teachers for Improving Education in the Classroom* (New York: Free Press, 1999).

169 **through work and fierce effort:** "Japanese Education Method Solves 21st Century Teaching Challenges in America," Japan society.org, Japan Society, February 26, 2009, https://www.japansociety.org/page/about/press/japanese_education_method_solves_21st_century_teaching_challenges_in_america.

169 **ubiquitous phrase *"good job!"*:** Jean M. Twenge and Elise C. Freeman, "Generational Differences in Young Adults' Life Goals, Concern for Others, and Civic Orientation, 1966–2009," PhD diss., San Diego State University, 2012.

172 **key attributes of confidence:** "Teenage Mental Health: What Helps and What Hurts?," 2009, http://www.dcya.gov.ie/documents/publications/MentalHealthWhatHelpsAndWhatHurts.pdf.

177 **to the breaking point:** Liz Funk, *Supergirls Speak Out: Inside the Secret Crisis of Overachieving Girls* (New York: Touchstone, 2009).

178 **figure for boys is 18 percent:** OECD, "Gender and Sustainable Development: Maximizing the Economic, Social and Environmental Role of Women," 2008, http://www.oecd.org/social/40881538.pdf.

182 **A 2011 Catalyst survey:** Catalyst.org, "The Promise of Future Leadership: Highly Talented Employees in the Pipeline," 2010, http://www.catalyst.org/knowledge/promise-future-leadership-research-program-highly-talented-employees-pipeline.

185 **do more harm than good:** Joanne V. Wood, Elaine Perunovic, and John W. Lee, "Positive Self-Statements: Power for Some, Peril for Others," National Institutes of Health, July 2009.

8: The Science and the Art

197 **everyone else, even the men:** Cecilia L. Ridgeway, "Interaction and the conservation of gender inequality: Considering employment," *American Sociological Review* (1997): 218–35.

198 **a recent Stanford University study:** A. O'Neill and Charles O'Reilly, "Overcoming the backlash effect: Self-Monitoring and Women's Promotions," *Journal of Occupational and Organizational Psychology*, 2011.

199 **company's mission, for example:** David A. Matsa et al., "A Female Style in Corporate Leadership? Evidence from Quotas," *American Economic Journal: Applied Economics*, forthcoming.

About the Authors

KATTY KAY is the Washington anchor for *BBC World News America*. She is a regular guest on NBC's *Meet the Press* and MSNBC's *Morning Joe*. She lives in Washington, DC, with her husband and four children.

CLAIRE SHIPMAN is a correspondent for ABC News and *Good Morning America*, covering politics, international affairs, and women's issues. She lives in Washington, DC, with her husband, two children, and a new puppy.